Artificial Neural Networks for Engineers and Scientists

Solving Ordinary Differential Equations

Artificial Neural Networks for Engineers and Scientists

Solving Ordinary Differential Equations

Snehashish Chakraverty and Susmita Mall

CRC Press is an imprint of the
Taylor & Francis Group, an **informa** business

CRC Press
Taylor & Francis Group
6000 Broken Sound Parkway NW, Suite 300
Boca Raton, FL 33487-2742

Printed on acid-free paper

International Standard Book Number-13: 978-1-4987-8138-1 (Hardback)

Library of Congress Cataloging-in-Publication Data

Names: Chakraverty, Snehashish. | Mall, Susmita.
Title: Artificial neural networks for engineers and scientists : solving ordinary differential equations / Snehashish Chakraverty & Susmita Mall.
Description: Boca Raton : CRC Press, 2017. | Includes bibliographical references.
Identifiers: LCCN 2017004411| ISBN 9781498781381 (hardback) | ISBN 9781498781404 (ebook)
Subjects: LCSH: Differential equations--Data processing. | Engineering mathematics--Data processing. | Artificial intelligence.
Classification: LCC QA372 .C42527 2017 | DDC 515/.3520285632--dc23
LC record available at https://lccn.loc.gov/2017004411

Visit the Taylor & Francis Web site at
http://www.taylorandfrancis.com

and the CRC Press Web site at
http://www.crcpress.com

Contents

Preface

Differential equations play a vital role in various fields of science and engineering. Many real-world problems in engineering, mathematics, physics, chemistry, economics, psychology, defense, etc., may be modeled by ordinary or partial differential equations. In most of the cases, an analytical/ exact solution of differential equations may not be obtained easily. Therefore, various types of numerical techniques have been developed by researchers to solve such equations, such as Euler, Runge–Kutta, predictor-corrector, finite difference, finite element, and finite volume techniques. Although these methods provide good approximations to the solution, they require the discretization of the domain into a number of finite points/elements, in general. These methods provide solution values at predefined points, and computational complexity increases with the number of sampling points.

In recent decades, among the various machine intelligence procedures, artificial neural network (ANN) methods have been established as powerful techniques to solve a variety of real-world problems because of ANN's excellent learning capacity. ANN is a computational model or an information processing paradigm inspired by the biological nervous system. Recently, a lot of attention has been devoted to the study of ANN for solving differential equations. The approximate solution of differential equations by ANN is found to be advantageous but it depends upon the ANN model that one considers. Here, our target is to handle ordinary differential equations (ODEs) using ANN. The traditional numerical methods are usually iterative in nature, where we fix the step size before the start of the computation. After the solution is obtained, if we want to know the solution between steps, the procedure needs to be repeated from the initial stage. ANN may be one of the ways to overcome this repetition of iterations.

In solving differential equations by ANN, no desired values are known and the output of the model can be generated by training only. As per the existing training algorithm, the architecture of a neural model is problem dependent and the number of nodes, etc., is taken by trial-and-error method where the training depends upon the weights of the connecting nodes. In general, these weights are taken as random numbers that dictate the training. This book includes recently developed new ANN models to handle various types of ODES. A brief outline of each chapter is given next.

The preliminaries of ANN are presented in Chapter 1. Definitions of ANN architecture, paradigm of learning, activation functions, leaning rules and learning processes, etc., are reviewed here. Chapter 2 describes preliminaries of ODEs. We recall the definitions that are relevant to the present book such as linear ODEs, nonlinear ODEs, initial and boundary value problems, etc. Chapter 3 deals with traditional multilayer ANN models to solve first- and higher-order ODEs. In the training algorithm, the number of nodes in the hidden layer is chosen by a trial-and-error method. The initial weights are random numbers as per the desired number of nodes. A simple feed-forward neural network and an unsupervised error back-propagation algorithm have been used. The chapter also addresses the general formulation of ODEs using multilayer ANN, formulation of *n*th-order initial value as well as boundary value problems, system of ODEs, and computation of gradient.

In Chapter 4, the recently developed regression-based neural network (RBNN) model is used to handle ODEs. In the RBNN model, the number of nodes in the hidden layer may be fixed according to the degree of polynomial in the regression. The coefficients involved are taken as initial weights to start with the neural training. A variety of first- and higher-order ODEs are solved as example problems. A single-layer functional link artificial neural network (FLANN) is presented in Chapter 5. In FLANN, the hidden layer is replaced by a functional expansion block for enhancement of the input patterns using orthogonal polynomials such as Chebyshev, Legendre, and Hermite. In this chapter, we discuss various types of single-layer FLANNs such as Chebyshev neural network (ChNN), Legendre neural network (LeNN), simple orthogonal polynomial–based neural network (SOPNN), and Hermite neural network (HeNN) used to solve linear and nonlinear ODEs. Chapter 6 introduces a single-layer FLANN model with regression-based weights to solve initial value problems. A second-order singular nonlinear initial value problem, namely, a Lane–Emden equation, has been solved using multi- and single-layer ANNs in Chapter 7. Next, Chapter 8 addresses the Emden–Fowler equations and their solution using multilayer ANN and single-layer FLANN methods.

Duffing oscillator equations have been considered in Chapter 9. Single-layer FLANN models, namely, SOPNN and HeNN models, have been used in this chapter to handle Duffing oscillator equations. Finally, Chapter 10 presents an ANN solution to the nonlinear van der Pol–Duffing oscillator equation where single-layer SOPNN and HeNN methods have been employed.

This book aims to provide a systematic understanding of ANN along with the new developments in ANN training. A variety of differential equations are solved here using the methods described in each chapter to show their reliability, powerfulness, and easy computer implementation. The book provides comprehensive results and up-to-date and self-contained review of the topic along with an application-oriented treatment of the use of newly developed ANN methods to solve various types of differential equations. It may be worth mentioning that the methods presented may very well be adapted for use in various other science and engineering disciplines where problems are modeled as differential equations and when exact and/or other traditional numerical methods may not be suitable. As such, this book will certainly prove to be essential for students, scholars, practitioners, researchers, and academicians in the assorted fields of engineering and sciences interested in modeling physical problems with ease.

Snehashish Chakraverty
Susmita Mall
National Institute of Technology, Rourkela

Acknowledgments

We express our sincere appreciation and gratitude to those outstanding people who have supported and happened to be the continuous source of inspiration throughout this book writing project.

The first author expresses his sincere gratitude to his parents, who always inspired him during the writing of this book. He also thanks his wife, Shewli Chakraborty, and his daughters, Shreyati and Susprihaa, for their continuous motivation. His PhD students and the NIT Rourkela facilities were also an important source of support in completing this book.

The second author expresses sincere gratitude and special appreciation to her father Sunil Kanti Mall, mother Snehalata Mall, father-in-law Gobinda Chandra Sahoo, mother-in-law Rama Mani Sahoo, and family members for their help, motivation, and invaluable advice. Writing this book would not have been possible without the love, unconditional support, and encouragement of her husband Srinath Sahoo and daughter Saswati.

The second author of this book conveys her heartfelt thanks to the family members of the first author, especially to his wife and daughters for their love, support, and inspiration.

Also, the authors gratefully acknowledge all the contributors and the authors of the books and journals/conferences papers listed in the book. Finally, they greatly appreciate the efforts of the entire editorial team of the publisher for their continuous support and help.

Snehashish Chakraverty
Susmita Mall

Authors

Dr. Snehashish Chakraverty is a professor of mathematics at National Institute of Technology, Rourkela, Odisha, India. He earned his PhD from the Indian Institute of Technology—Roorkee in 1992, and did postdoctoral research at ISVR, the University of Southampton, UK, and at Concordia University, Canada. He was a visiting professor at Concordia University and at McGill University, in Canada, and also at the University of Johannesburg, South Africa. Dr. Chakraverty has authored ten books, including two that will be published in 2017, and has published more than 250 research papers in journals and for conferences. He has served as president of the Section of Mathematical Sciences (including Statistics) of the Indian Science Congress (2015–2016) and as the vice president of the Orissa Mathematical Society (2011–2013). Dr. Chakraverty has received several prestigious awards, including the INSA International Bilateral Exchange Program, Platinum Jubilee ISCA Lecture, CSIR Young Scientist, BOYSCAST, UCOST Young Scientist, Golden Jubilee CBRI Director's Award, and the Roorkee University Gold Medals. He has undertaken around 16 major research projects as principle investigator, funded by different agencies. His present research areas include soft computing, numerical analysis, differential equations, vibration and inverse vibration problems, and mathematical computation.

Dr. Susmita Mall received her PhD from the National Institute of Technology, Rourkela, Odisha, India, in 2016 and MSc in mathematics from Ravenshaw University, Cuttack, Odisha. She was awarded the Women Scientist Scheme-A (WOS-A) fellowship under the Department of Science and Technology (DST), Government of India, to pursue her PhD studies. She has published ten research papers in international refereed journals and five in conferences. Her current research interest includes mathematical modeling, artificial neural network, differential equations, and numerical analysis.

Reviewers

Rama Bhata
Mechanical Engineering
Concordia University
Montreal, Quebec, Canada

Fernando Buarque
Polytechnic School of Pernambuco
University of Pernambuco
Pernambuco, Brazil

T.R. Gulati
Department of Mathematics
Indian Institute of Technology Roorkee
Roorkee, Uttarakhand, India

Aliakbar Montazer Haghighi
Department of Mathematics
Prairie View A&M University
Prairie View, Texas

Nikos D. Lagaros
Institute of Structural Analysis & Seismic Research
National Technical University
Athens, Greece

Tshilidzi Marwala
Department of Electrical and Electronic Engineering
University of Johannesburg
Johannesburg, South Africa

Sushmita Mitra
Machine Intelligence Unit
Indian Statistical Institute
Kolkata, West Bengal, India

A. Sahu
Mathematics & Computer Science
Coppin State University
Baltimore, Maryland

Tanmoy Som
Department of Mathematical Science
Indian Institute of Technology (BHU)
Varanasi, Uttar Pradesh, India

A.M. Wazwaz
Department of Mathematics
Saint Xavier University
Chicago, Illinois

1

Preliminaries of Artificial Neural Network

This chapter addresses basics of artificial neural network (ANN) architecture, paradigms of learning, activation functions, and leaning rules.

1.1 Introduction

Artificial neural network (ANN) is one of the popular areas of artificial intelligence (AI) research and also an abstract computational model based on the organizational structure of the human brain [1]. The simplest definition of ANN is provided by the inventor of one of the first neurocomputers, Dr. Robert Hecht-Nielsen. He defines a neural network as

> a computing system made up of a number of simple, highly interconnected processing elements, which process information by their dynamic state response to external inputs.

ANNs are processing devices (algorithms) that are loosely modeled after the neuronal structure of the mammalian cerebral cortex but on much smaller scales. Computer scientists have always been inspired by the human brain. In 1943, Warren S. McCulloch, a neuroscientist, and Walter Pitts, a logician, developed the first conceptual model of an ANN [1]. In their paper, they describe the concept of a neuron, a single cell living in a network of cells that receives inputs, processes those inputs, and generates an output.

ANN is a data modeling tool that depends upon various parameters and learning methods [2–8]. Neural networks are typically organized in layers. Layers are made up of a number of interconnected "neurons/nodes," which contain "activation functions." ANN processes information through neurons/nodes in a parallel manner to solve specific problems. ANN acquires knowledge through learning, and this knowledge is stored within interneuron connections' strength, which is expressed by numerical values called "weights." These weights are used to compute output signal values

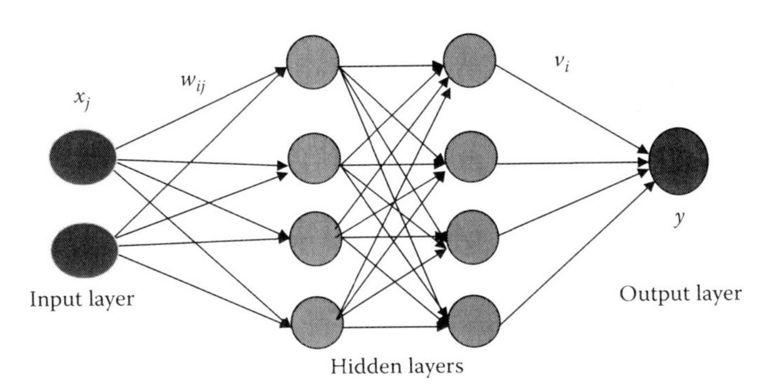

FIGURE 1.1
Structure of artificial neural network.

for a new testing input signal value. Patterns are presented to the network via the "input layer," which communicates to one or more "hidden layers," where the actual processing is done via a system of weighted "connections." The hidden layers then link to an "output layer," where the answer is output, as shown in Figure 1.1.

Here, x_j are input nodes, w_{ij} are weights from the input layer to the hidden layer, and v_i and y denote the weights from the hidden layer to the output layer and the output node, respectively.

The ANN method has been established as a powerful technique to solve a variety of real-world problems because of its excellent learning capacity [9–12]. This method has been successfully applied in various fields [8,13–25] such as function approximation, clustering, prediction, identification, pattern recognition, solving ordinary and partial differential equations, etc.

1.2 Architecture of ANN

It is a technique that seeks to build an intelligent program using models that simulate the working of neurons in the human brain. The key element of the network is the structure of the information processing system. ANN processes information in a similar way as the human brain does. The network is composed of a large number of highly interconnected processing elements (neurons) working in parallel to solve a specific problem.

Neural computing is a mathematical model inspired by the biological model. This computing system is made up of a large number of artificial neurons and a still larger number of interconnections among them. According to the structure of these interconnections, different classes of neural network architecture can be identified, as discussed next.

1.2.1 Feed-Forward Neural Network

In a feed-forward neural network, neurons are organized in the form of layers. Neurons in a layer receive input from the previous layer and feed their output to the next layer. Network connections to the same or previous layers are not allowed. Here, the data goes from the input node to the output node in a strictly feed-forward way. There is no feedback (back loops); that is, the output of any layer does not affect the same layer. Figure 1.2 shows the block diagram of the feed-forward ANN [8].

Here, $X=(x_1, x_2, \ldots, x_n)$ denotes the input vector, f is the activation function, and $O=(o_1, o_2, \ldots, o_m)$ is the output vector. Symbol $W=w_{ij}$ denotes the weight matrix or connection matrix, and $[WX]$ is the net input value, which is a scalar product of input vectors and weight vectors w_{ij}

$$W = \begin{bmatrix} w_{11} & w_{12} & \cdots & w_{1n} \\ w_{21} & w_{22} & \cdots & w_{2n} \\ \vdots & \vdots & \cdots & \vdots \\ w_{m1} & w_{m2} & \cdots & w_{mn} \end{bmatrix}$$

1.2.2 Feedback Neural Network

In a feedback neural network, the output of one layer routes back to the previous layer. This network can have signals traveling in both directions by the introduction of loops in the network. This network is very powerful and, at times, gets extremely complicated. All possible connections between neurons are allowed. Feedback neural networks are used in optimization problems, where the network looks for the best arrangement of interconnected factors. They are dynamic and their state changes continuously until they reach an equilibrium point (Figure 1.3).

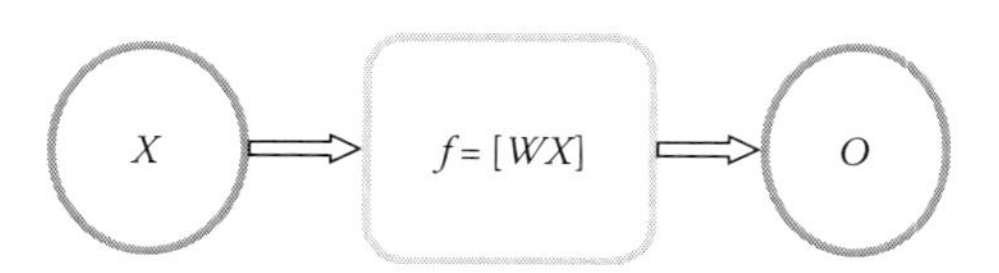

FIGURE 1.2
Block diagram of feed-forward ANN.

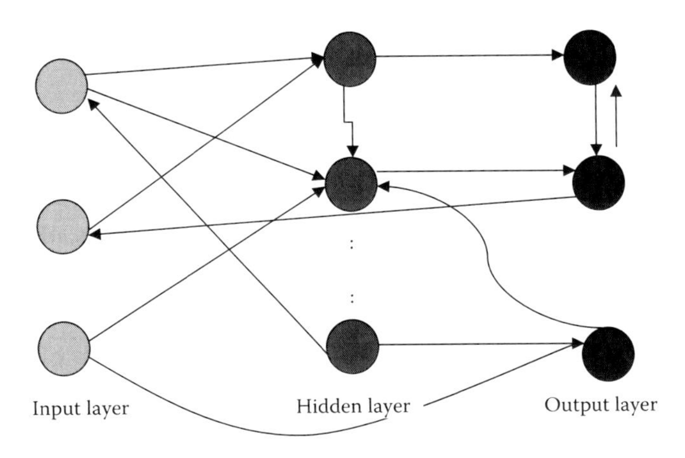

FIGURE 1.3
Diagram of feedback neural network.

1.3 Paradigms of Learning

The ability to learn and generalize from a set of training data is one of the most powerful features of ANN. The learning situations in neural networks may be classified into two types, namely, supervised and unsupervised.

1.3.1 Supervised Learning or Associative Learning

In supervised training, both inputs and outputs are provided. The network then processes the inputs and compares its resulting outputs against the desired outputs. A comparison is made between the network's computed output and the corrected expected output to determine the error. The error can then be used to change network parameters, which results in an improvement in performance. In other words, inputs are assumed to be at the beginning and outputs at the end of the causal chain. Models can include mediating variables between inputs and outputs.

1.3.2 Unsupervised or Self-Organization Learning

The other type of training is called unsupervised training. In unsupervised training, the network is provided with inputs but not with desired outputs. The system itself must then decide what features it will use to group the input data. This is often referred to as self-organization or adaptation. Unsupervised learning seems much harder than supervised learning, and this type of training generally fits into the decision problem framework because the goal is not to produce a classification but to make decisions that maximize rewards. An unsupervised learning task is to try to find the hidden structure in unlabeled data. Since the examples given to the learner are unlabeled, there is no error or reward signal to evaluate a potential solution.

1.4 Learning Rules or Learning Processes

Learning is the most important characteristic of the ANN model. Every neural network possesses knowledge that is contained in the values of the connection weights. Most ANNs contain some form of "learning rule" that modifies the weights of the connections according to the input patterns that it is presented with. Although there are various kinds of learning rules used by neural networks, the delta learning rule is often utilized by the most common class of ANNs called back-propagation neural networks (BPNNs). Back propagation is an abbreviation for the backward propagation of error.

There are various types of learning rules for ANN [8,13] such as the following:

1. Error back-propagation learning algorithm or delta learning rule
2. Hebbian learning rule
3. Perceptron learning rule
4. Widrow–Hoff learning rule
5. Winner-Take-All learning rule, etc.

The details of the above-listed types of learning may be found in any standard ANN books. As such, here, we will discuss only a bit about the first one because the same has been used in most of the investigations in this book.

1.4.1 Error Back-Propagation Learning Algorithm or Delta Learning Rule

Error back-propagation learning algorithm has been introduced by Rumelhart et al. [3]. It is also known as the delta learning rule [5,8,14] and is one of the most commonly used learning rules. This learning algorithm is valid for continuous activation function and is used in the supervised/unsupervised training method.

A simple perceptron can handle linearly separable or linearly independent problems. Taking the partial derivative of error of the network with respect to each of its weights, we can know the flow of error direction in the network. If we take the negative derivative and then proceed to add it to the weights, the error will decrease until it approaches a local minimum. We have to add a negative value to the weight or the reverse if the derivative is negative. Then, these partial derivatives are applied to each of the weights, starting from the

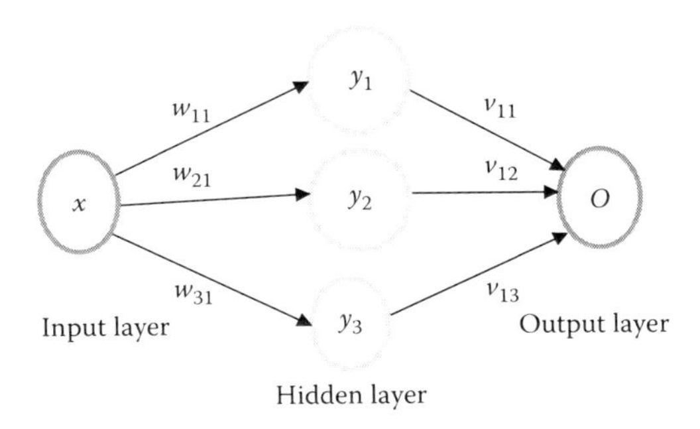

FIGURE 1.4
Architecture of multilayer feed forward neural network.

output layer weights to the hidden layer weights, and then from the hidden layer weights to the input layer weights.

In general, the training of the network involves feeding samples as input vectors, calculating the error of the output layer, and then adjusting the weights of the network to minimize the error. The average of all the squared errors E for the outputs is computed to make the derivative simpler. After the error is computed, the weights can be updated one by one. The descent depends on the gradient ∇E for the training of the network.

Let us consider a multilayer neural architecture containing one input node x, three nodes in the hidden layer y_j, $j = 1, 2, 3$, and one output node O. Now by applying feed-forward recall with error back-propagation learning to the model presented in Figure 1.4, we have the following algorithm [8]:

Step 1: Initialize the weights W from the input layer to the hidden layer and weights V from the hidden layer to the output layer. Choose the learning parameter η (lies between 0 and 1) and error $E_{\max}$.

Next, initially, error is taken as $E = 0$.

Step 2: Training steps start here.

Outputs of the hidden layer and the output layer are computed as follows:

$$y_j \leftarrow f\left(w_j x\right), \quad j = 1,2,3$$

$$o_k \leftarrow f\left(v_k y\right), \quad k = 1$$

where

w_j is the jth row of W for $j = 1, 2, 3$

v_k is the kth row of V for $k = 1$

f is the activation function

Step 3: Error value is computed as

$$E = \frac{1}{2}\left(d_k - o_k\right)^2 + E$$

Here,

d_k is the desired output

o_k is the output of ANN

Step 4: The error signal terms of the output layer and the hidden layer are computed as

$$\delta_{ok} = \left[\left(d_k - o_k\right) f'\left(v_k y\right)\right] \quad \text{(error signal of the output layer)}$$

$$\delta_{yj} = \left[\left(1 - y_j\right) f\left(w_j x\right)\right] \delta_{ok} v_{kj} \quad \text{(error signal of the hidden layer)}$$

where $o_k = f(v_k y)$, $j = 1, 2, 3,$ and $k = 1$.

Step 5: Compute components of error gradient vectors as

$\partial E / \partial w_{ji} = \delta_{yj} x_i$ for j = 1, 2, 3 and i = 1. (For the particular ANN model, see Figure 1.4.)

$\partial E / \partial v_{kj} = \delta_{ok} y_j$ for j = 1, 2, 3 and k = 1. (For the particular ANN model, see Figure 1.4.)

Step 6: Weights are modified using the gradient descent method from the input layer to the hidden layer and from the hidden layer to the output layer as

$$w_{ji}^{n+1} = w_{ji}^n + \Delta w_{ji}^n = w_{ji}^n + \left(-\eta \frac{\partial E}{\partial w_{ji}^n}\right)$$

$$v_{kj}^{n+1} = v_{kj}^n + \Delta v_{kj}^n = v_{kj}^n + \left(-\eta \frac{\partial E}{\partial v_{kj}^n}\right)$$

where

η is the learning parameter

n is the iteration step

E is the error function

Step 7: If $E = E_{\max}$, terminate the training session; otherwise, go to step 2 with $E \leftarrow 0$ and initiate a new training.

The generalized delta learning rule propagates the error back by one layer, allowing the same process to be repeated for every layer.

1.5 Activation Functions

An activation or transfer function is a function that acts upon the net (input) to get the output of the network. It translates input signals to output signals.

It acts as a squashing function such that the output of the neural network lies between certain values (usually between 0 and 1, or −1 and 1).

Five types of activation functions are commonly used:

1. Unit step (threshold) function
2. Piecewise linear function
3. Gaussian function
4. Sigmoid function
   a. Unipolar sigmoid function
   b. Bipolar sigmoid function
5. Tangent hyperbolic function

Throughout this book, we have used sigmoid and tangent hyperbolic functions only, which are nonlinear, monotonic, and continuously differentiable.

1.5.1 Sigmoid Function

The sigmoid function is defined as a strictly increasing and continuously differentiable function.

It exhibits a graceful balance between linear and nonlinear behavior.

1.5.1.1 Unipolar Sigmoid Function

The unipolar sigmoid function is shown in Figure 1.5 and defined by the formula

$$f(x) = \frac{1}{1+e^{-\lambda x}}$$

where $\lambda > 0$ is the slope of the function.

The output of the uniploar sigmoid function lies in [0, 1].

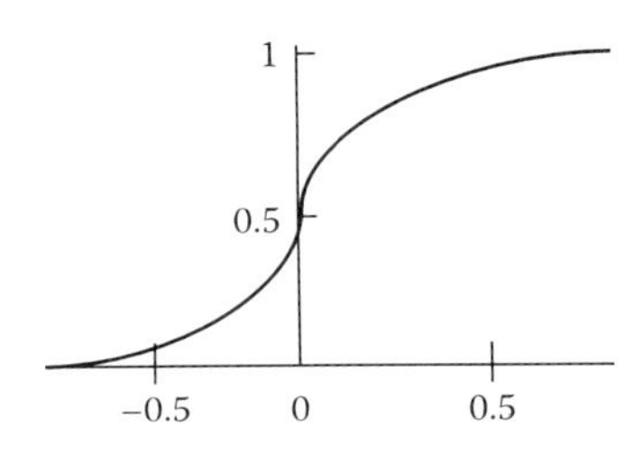

FIGURE 1.5
Plot of unipolar sigmoid function.

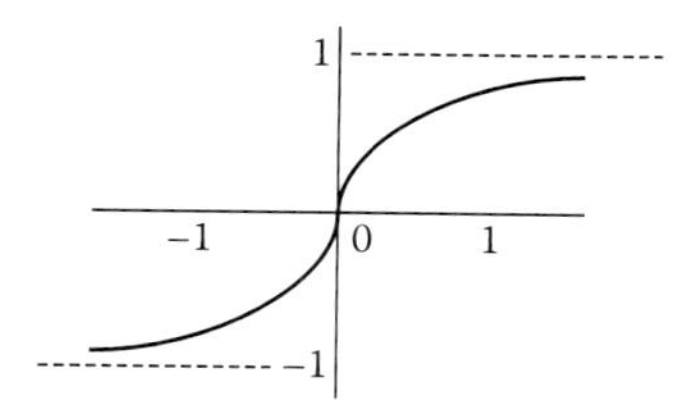

FIGURE 1.6
Plot of bipolar sigmoid function.

1.5.1.2 Bipolar Sigmoid Function

The bipolar sigmoid function is formulated as

$$f(x) = \frac{2}{1+e^{-\lambda x}} - 1 = \frac{1-e^{-\lambda x}}{1+e^{-\lambda x}}$$

The output of the bipolar sigmoid function lies between [−1, 1]. Figure 1.6 shows the plot of the function.

1.5.2 Tangent Hyperbolic Function

The tangent hyperbolic function is defined as

$$f(x) = \frac{e^{x} - e^{-x}}{e^{x} + e^{-x}} = \frac{e^{2x} - 1}{e^{2x} + 1}$$

The output of the tangent hyperbolic function lies in [−1, 1] and plot of tangent hyperbolic function is same as bipolar sigmoid function.

Further details of ANN architecture, paradigms of learning, learning algorithms, activation functions, etc., may be found in standard ANN books.

References

1. W.S. McCulloch and W. Pitts. A logical calculus of the ideas immanent in nervous activity. *Bulletin of Mathematical Biophysics*, 5(4): 115–133, December 1943.
2. M. Minsky and S. Papert. *Perceptrons*. MIT Press, Cambridge, MA, 1969.
3. D.E. Rumelhart, G.E. Hinton, and J.L. McClelland. *Parallel Distributed Processing*. MIT Press, Cambridge, MA, 1986.
4. R.P. Lippmann. An introduction to computing with neural nets. *IEEE ASSP Magazine*, 4: 4–22, 1987.
5. J.A. Freeman and D.M. Skapura. *Neural Networks: Algorithms, Applications, and Programming Techniques*. Addison-Wesley Publishing Company, Boston, MA, 1991.

6. J.A. Anderson. *An Introduction to Neural Networks*. MIT Press, Cambridge, London, England, U.K., 1995.
7. G.E. Hinton, S. Osindero, and Y.-W. Teh. A fast learning algorithm for deep belief nets. *Neural Computation*, 18: 1527–1554, 2006.
8. J.M. Zurada. *Introduction to Artificial Neural Systems*. West Publishing Co., St. Paul, MN, 1992.
9. K. Hornik, M. Stinchcombe, and H. White. Multilayer feed forward networks are universal approximators. *Neural Networks*, 2(5): 359–366, January 1989.
10. T. Khanna. *Foundations of Neural Networks*. Addison-Wesley Press, Boston, USA, 1990.
11. R.J. Schalkoff. *Artificial Neural Networks*. McGraw-Hill, New York, 1997.
12. P.D. Picton. *Neural Networks*, 2nd edn. Palgrave Macmillan, New York, 2000.
13. R. Rojas. *Neural Networks: A Systematic Introduction*. Springer, Berlin, Germany, 1996.
14. S.S. Haykin. *Neural Networks: A Comprehensive Foundation*. Prentice Hall Inc., Upper Saddle River, NJ, 1999.
15. B. Yegnanarayana. *Artificial Neural Networks*. Eastern Economy Edition, Prentice-Hall of India Pvt. Ltd., India, 1999.
16. M. Hajek. *Neural Network*. University of KwaZulu-Natal Press, Durban, South Africa, 2005.
17. D. Graupe. *Principles of Artificial Neural Networks*. World Scientific Publishing Co. Ltd., Toh Tuck Link, Singapore, 2007.
18. N.D. Lagaros and M. Papadrakakis. Learning improvement of neural networks used in structural optimization. *Advances in Engineering Software*, 35(1): 9–25, January 2004.
19. S. Chakraverty. Identification of structural parameters of two-storey shear building by the iterative training of neural networks. *Architectural Science Review Australia*, 50(4): 380–384, July 2007.
20. S. Mall and S. Chakraverty. Comparison of artificial neural network architecture in solving ordinary differential equations. *Advances in Artificial Neural Systems*, 2013: 1–24, October 2013.
21. S. Mall and S. Chakraverty. Regression-based neural network training for the solution of ordinary differential equations. *International Journal of Mathematical Modelling and Numerical Optimization*, 4(2): 136–149, 2013.
22. S. Chakraverty and S. Mall. Regression based weight generation algorithm in neural network for solution of initial and boundary value problems. *Neural Computing and Applications*, 25(3): 585–594, September 2014.
23. S. Mall and S. Chakraverty. Regression based neural network model for the solution of initial value problem. *National Conference on Computational and Applied Mathematics in Science and Engineering (CAMSE-2012)*, VNIT, Nagpur, India, December 2012.
24. S. Mall and S. Chakraverty. Chebyshev neural network based model for solving Lane–Emden type equations. *Applied Mathematics and Computation*, 247: 100–114, November 2014.
25. S. Mall and S. Chakraverty. Numerical solution of nonlinear singular initial value problems of Emden–Fowler type using Chebyshev neural network method. *Neurocomputing*, 149: 975–982, February 2015.

2

Preliminaries of Ordinary Differential Equations

It is well known that differential equations (DEs) are the backbone of physical systems. The mathematical representation of science and engineering problems is called a mathematical model. The corresponding equation for such physical systems is generally given by DEs and depends upon the particular physical problem. An equation involving one dependent variable and its derivative with respect to one or more independent variables is called a differential equation. It may be noted that most of the fundamental laws of nature can be formulated as DEs, namely, the growth of a population, motion of a satellite, flow of current in an electric circuit, change in prices of commodities, conduction of heat in a rod, vibration of structures, etc.

Let us consider $y=f(x)$; that is, y is a function of x, and its derivative dy/dx can be interpreted as the rate of change of y (dependent variable) with respect to x (independent variable). In any natural process, the variables involved and their rate of change are connected with one another by means of the basic scientific principles that govern the process. The concepts and details of differential equations may be found in many standard books [1–14] and thus are not repeated here. But a few important and basic points related to DEs are discussed in Section 2.1, which may help readers to model them by artificial neural network (ANN), which happens to be the main focus of this book.

The following are some examples of DEs:

$$\frac{dy}{dx} + 2xy = e^{-x^2} \tag{2.1}$$

$$\frac{d^3y}{dx^3} - 3\frac{dy}{dx} + 2 = 4\tan x \tag{2.2}$$

$$\left(\frac{d^2y}{dx^2}\right)^2 - \left(\frac{dy}{dx}\right)^4 + y = 0 \tag{2.3}$$

$$\frac{d^2y}{dx^2} + \frac{2}{x}\frac{dy}{dx} + y^5 = 0 \tag{2.4}$$

$$\frac{\partial v}{\partial x}+\left(\frac{\partial v}{\partial y}\right)^3=0 \tag{2.5}$$

$$\frac{\partial^2 v}{\partial x^2}+\frac{\partial^2 v}{\partial y^2}+\frac{\partial^2 v}{\partial z^2}=-\frac{1}{2v}\frac{\partial v}{\partial x} \tag{2.6}$$

$$\frac{\partial^2 v}{\partial x^2}+\frac{\partial^2 v}{\partial y^2}\frac{\partial v}{\partial x}+\frac{\partial^2 v}{\partial z^2}=0 \tag{2.7}$$

2.1 Definitions

2.1.1 Order and Degree of DEs

The order of a differential equation is defined as the order of the highest derivative involved in the given DE. The degree of a differential equation is the highest power (positive integral index) of the highest-order derivative of the differential equation.

In the previous examples, Equations 2.1 and 2.5 are first-order DEs; Equations 2.3, 2.4, 2.6, and 2.7 are second-order DEs; and Equation 2.2 is a third-order DE. Similarly, Equations 2.1, 2.2, 2.4, 2.6, and 2.7 are first-degree DEs, Equation 2.3 is a second-degree DE, and Equation 2.5 is a third-degree DE.

2.1.2 Ordinary Differential Equation

Ordinary differential equation (ODE) is a differential equation that contains only one independent variable so that all the derivatives occurring in it are ordinary derivatives. Equations 2.1 through 2.4 are examples of ODEs.

The general form of an nth-order ODE is

$$f\left(x, y, \frac{dy}{dx}, \frac{d^2y}{dx^2}, \ldots, \frac{d^n y}{dx^n}\right)=0$$

or

$$f\left(x, y, y', y'', y''', \ldots, y^n\right)=0$$

2.1.3 Partial Differential Equation

Partial differential equation (PDE) is a differential equation that contains more than one independent variable and their partial derivatives. PDEs are equations that involve rates of change with respect to continuous variables.

PDEs describe various phenomena such as sound, heat, electrostatics, electrodynamics, fluid flow, elasticity, etc.

In the previous examples, Equations 2.5 through 2.7 are PDEs.

A PDE for the function $v(x_1, x_2, \ldots, x_n)$ is an equation of the form

$$f\left(x_1, x_2, \ldots, x_n, \frac{\partial v}{\partial x_1}, \frac{\partial v}{\partial x_2}, \ldots, \frac{\partial v}{\partial x_n}, \frac{\partial^2 v}{\partial x_1 \partial x_2}, \ldots, \frac{\partial^2 v}{\partial x_1 \partial x_n}, \ldots\right) = 0$$

2.1.4 Linear and Nonlinear Differential Equations

Linear differential equations are those in which (1) the dependent variable and its derivatives appear only in first degree, and (2) products of derivatives and the dependent variable do not occur.

Linear DEs can be expressed as

$$c_0 \frac{d^n y}{dx^n} + c_1 \frac{d^{n-1} y}{dx^{n-1}} + \cdots + c_{n-1} \frac{dy}{dx} + c_n y = f(x)$$

where the coefficient $c_0, \ldots, c_n$ and $f(x)$ denote constants and the functions of x, respectively.

The following are some examples of linear DEs:

$$\frac{dy}{dx} + xy = \sin x$$

$$y'' - 4xy' + \left(4x^2 - 1\right) y = -3e^{x^2}$$

Differential equations that are not linear are called nonlinear differential equations; that is, the products of the unknown function and derivatives are allowed and degree of dependent variable is >1.

The following are some examples of nonlinear DEs:

$$\frac{d^2 y}{dx^2} + \frac{2}{x} y \frac{dy}{dx} + e^{-x} = 0$$

$$\frac{d^2 x}{dt^2} + \alpha \frac{dx}{dt} + \beta x + \gamma x^3 = F \cos \omega t$$

2.1.5 Initial Value Problem

Many problems in science and engineering can be modeled by ODEs or PDEs, along with one or more supplementary conditions. If these conditions are given at one point of independent variable, the problem is called initial value problem (IVP) and these conditions are known as initial conditions.

Let us consider a first-order ODE

$$\frac{dy}{dx} = f(x,y)$$

which satisfies the initial condition $y(x_0) = y_0$.

Similarly, for a second-order ODE

$$\frac{d^2y}{dx^2} = f\left(x, y, \frac{dy}{dx}\right)$$

subject to initial conditions $y(x_0) = y_0$, $y'(x_0) = y_1$, an nth-order IVP with n initial conditions may be written as

$$\frac{d^n y}{dx^n} = f\left(x, y(x), \frac{dy}{dx}, \frac{d^2y}{dx^2}, \ldots, \frac{d^{n-1}y}{dx^{n-1}}\right)$$

with initial conditions

$$\begin{aligned} y(x_0) &= y_0, \\ y'(x_0) &= y_1, \\ y''(x_0) &= y_2, \\ &\vdots \\ &\vdots \\ y^{n-1}(x_0) &= y_{n-1}. \end{aligned}$$

2.1.6 Boundary Value Problem

Boundary value problem (BVP) is a differential equation together with a set of additional conditions, called the boundary conditions. The boundary conditions are specified at the extreme (boundaries) points or at more than one point of the independent variable.

Here, we have considered a second-order ODE in the interval (x_0, x_1)

$$y''(x) = f(x,y)$$

Let $y(x)$ is the solution of the BVP, which satisfies one of the following pairs of boundary conditions:

$$y(x_0) = y_0, \quad y(x_1) = y_1$$

$$y(x_0) = y_0, \quad y'(x_1) = y_1$$

$$y'(x_0) = y_0, \quad y(x_1) = y_1$$

where y_0 and y_1 are constants.

References

1. W.E. Boyce and R.C. Diprima. *Elementary Differential Equations and Boundary Value Problems*. John Wiley & Sons, Inc., New York, 2001.
2. Y. Pinchover and J. Rubinsteinan. *Introduction to Partial Differential Equations*. Cambridge University Press, Cambridge, England, 2005.
3. D.W. Jordan and P. Smith. *Nonlinear Ordinary Differential Equations: Problems and Solutions*. Oxford University Press, Oxford, U.K., 2007.
4. R.P. Agrawal and D. O'Regan. *An Introduction to Ordinary Differential Equations*. Springer, New York, 2008.
5. D.G. Zill and M.R. Cullen. *Differential Equations with Boundary Value Problems*. Brooks/Cole, Cengage Learning, Belmont, CA, 2008.
6. H.J. Ricardo. *A Modern Introduction to Differential Equations*, 2nd edn. Elsevier Academic Press, CA, 2009.
7. M.V. Soare, P.P. Teodorescu, and I. Toma. *Ordinary Differential Equations with Applications to Mechanics*. Springer, New York, 2007.
8. A.D. Poyanin and V.F. Zaitsev. *Handbook of Exact Solutions for Ordinary Differential Equations*. Chapman & Hall/CRC, Boca Raton, FL, 2000.
9. J. Douglas and B.F. Jones. Predictor-Corrector methods for nonlinear parabolic differential equations. *Journal for Industrial and Applied Mathematics*, 11(1): 195–204, March 1963.
10. G.F. Simmons. *Differential Equations with Applications and Historical Notes*. Tata McGraw-Hill, Inc., New Delhi, India, 1972.
11. J. Sinharoy and S. Padhy. *A Course on Ordinary and Partial Differential Equations with Applications*. Kalyani Publishers, New Delhi, India, 1986.
12. A. Wambecq. Rational Runge–Kutta methods for solving systems of ordinary differential equations. *Computing*, 20(4): 333–342, December 1978.
13. R. Norberg. Differential equations for moments of present values in life insurance. *Insurance: Mathematics and Economics*, 17(2): 171–180, October 1995.
14. C.J. Budd and A. Iserles. Geometric integration: Numerical solution of differential equations on manifolds. *Philosophical Transactions: Royal Society*, 357: 945–956, November 1999.

3

Multilayer Artificial Neural Network

In this chapter, the traditional multilayer artificial neural network (ANN) model has been used for solving ordinary differential equations (ODEs). We have considered a multilayer ANN model with one input layer containing a single node, a hidden layer with m nodes, and one output node. The ANN trial solution of the differential equation is a sum of two terms. The first term satisfies the initial or boundary conditions, while the second term contains the output of the ANN with adjustable parameters. The feed-forward neural network model and unsupervised error back-propagation algorithm have been used. Modification of network parameters has been done without the use of any optimization technique.

In recent years, the ANN model has been widely applied to a variety of real-world scenarios [1,2]. Accordingly, we have used here the ANN model for solving ODEs. The ANN-based solution has many advantages compared with other traditional numerical methods. The approximate solution is continuous over the domain of integration. Computational complexity does not increase considerably with the number of sampling points. Moreover, traditional numerical methods are usually iterative in nature, where we fix the step size before the start of the computation. After the solution is obtained, if we want to know the solution in between steps, then again the procedure is to be repeated from the initial stage. ANN may be one of the ways where we may overcome this repetition of iterations. Also, we may use it as a black box to get numerical results at any arbitrary point in the domain after training of the model. In the following, we mention a few investigations where ANNs are utilized for solving differential equations.

The Hopfield neural network model for solving ODEs has been introduced by Lee and Kang [3]. Meade and Fernandez [4,5] investigated linear and nonlinear ODEs using the feed-forward neural network model and B_1 splines. Lagaris et al. [6] proposed multilayer ANN architecture along with an optimization technique, to solve both ordinary and partial differential equations. Kumar and Yadav [7] surveyed multilayer perceptrons and radial basis function neural network methods for the solution of differential equations. Ibraheem and Khalaf [8] solved boundary value problems (BVPs) using the multilayer neural network method. Liu and Jammes [9] presented a procedure based on neural networks to solve ODEs. Mall and Chakraverty [10,11] proposed a regression-based multilayer neural network model for solving initial and boundary value problems. Malek and Beidokhti [12] presented a multilayer neural network with an optimization technique to solve differential equations of lower as well as of higher order. Yazdi et al. [13] proposed the kernel least mean square algorithm for solving first- and second-order ODEs. Tsoulos and Lagaris [14] utilized feed-forward neural networks, grammatical evolution, and a local optimization procedure to solve ordinary and partial differential equations and a system of ODEs. In another work, Parisi et al. [15], a steady-state heat transfer problem has been solved by using unsupervised ANN.

3.1 Structure of Multilayer ANN Model

We have considered a three-layer ANN model for the present problem. Figure 3.1 shows the structure of ANN, which consists of a single input node (x) along with biases (u_j), a hidden layer, and a single output layer consisting of one output node. Initial weights from the input layer to the hidden layer (w_j) and from the hidden layer to the output layer (v_j) are taken as arbitrary (random), and the number of nodes in the hidden layer are considered by the trial-and-error method.

3.2 Formulations and Learning Algorithm of Multilayer ANN Model

3.2.1 General Formulation of ODEs Based on ANN Model

In recent years, several methods have been proposed to solve ordinary as well as partial differential equations. First, we consider a general form of differential equations that represents ODEs [6]

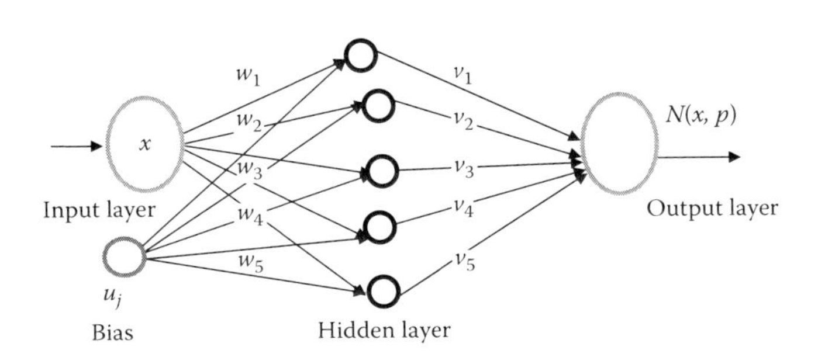

FIGURE 3.1
Structure of multilayer ANN.

$$G\left(x, y(x), \nabla y(x), \nabla^2 y(x), \ldots, \nabla^n y(x)\right) = 0, \quad x \in \bar{D} \subseteq R \tag{3.1}$$

where

G is a function that defines the structure of a differential equation
$y(x)$ denotes the solution
∇ is the differential operator
$\bar{D}$ is the discretized domain over a finite set of points

One may note that $x \in \bar{D} \subset R$ for ODEs. Let $y_t(x,p)$ denote the ANN trial solution for ODEs with adjustable parameters p (weights and biases); then, the this general differential equation changes to the form

$$G\left(x, y_t(x,p), \nabla y_t(x,p), \nabla^2 y_t(x,p), \ldots, \nabla^n y_t(x,p)\right) = 0 \tag{3.2}$$

The trial solution $y_t(x,p)$ of feed-forward neural network with input x and parameters p may be written in the form [6]

$$y_t(x,p) = A(x) + F\left(x, N(x,p)\right) \tag{3.3}$$

where $A(x)$ satisfies the initial or boundary conditions and contains no adjustable parameters, whereas $N(x,p)$ is the output of feed-forward neural network with parameters p and input x. The second term $F(x,N(x,p))$ makes no contribution to the initial/boundary conditions but is the output of the neural network model whose weights and biases are adjusted to minimize the error function to obtain the final ANN solution, $y_t(x,p)$.

Here, the network output $N(x,p)$ is formulated as

$$N(x,p) = \sum_{j=1}^{m} v_j s(z_j) \tag{3.4}$$

where $z_j = w_j x + u_j$, w_j denotes the weight from the input unit to the hidden unit j, v_j denotes the weight from the hidden unit j to the output unit, u_j are the biases, and $s(z_j)$ is the activation function (sigmoid, tangent hyperbolic, etc.).

It may be noted that in the training method, we start with the given weights and biases and train the model to modify the weights in the given domain of the problem. In this procedure, our aim is to minimize the error function. Accordingly, we include the formulation of the error function for initial value problems (IVPs) next.

3.2.2 Formulation of *n*th-Order IVPs

Let us consider a general nth-order IVP [12]

$$\frac{d^n y}{dx^n} = f\left(x, y, \frac{dy}{dx}, \frac{d^2 y}{dx^2}, \dots, \frac{d^{n-1} y}{dx^{n-1}}\right) \quad x \in [a,b] \tag{3.5}$$

with initial conditions $y^{(i)}(a) = A_i$, $i = 0, 1, \dots, n-1$.

The corresponding ANN trial solution may be constructed as

$$y_t(x,p) = \sum_{i=0}^{n-1} u_i x^i + (x-a)^n N(x,p) \tag{3.6}$$

where $\{u_i\}_{i=0}^{n-1}$ are the solutions to the upper triangle system of n linear equations in the form [12]

$$\sum_{i=j}^{n-1} \binom{j}{i} j! a^{i-j} u_i = A_j, \quad j = 0, 1, 2, \dots, n-1$$

The general formula of the error function for ODEs may be written as follows:

$$E(x,p) = \sum_{i=1}^{h} \frac{1}{2}\left(\frac{d^n y_t(x_i,p)}{dx^n} - f\left[x_i, y_t(x_i,p), \frac{dy_t(x_i,p)}{dx}, \dots, \frac{d^{n-1} y_t(x_i,p)}{dx^{n-1}}\right]\right)^2 \tag{3.7}$$

It may be noted that the multilayer ANN model is considered with one input node x (having h number of data) and a single output node $N(x,p)$ for ODEs.

Here, an unsupervised error back-propagation algorithm is used for minimizing the error function. In order to update the network parameters (weights and biases) from the input layer to the hidden layer and from the hidden layer to the output layer, we use the following expressions [15]:

$$w_j^{k+1} = w_j^k + \Delta w_j^k = w_j^k + \left(-\eta \frac{\partial E(x,p)^k}{\partial w_j^k}\right) \tag{3.8}$$

$$v_j^{k+1} = v_j^k + \Delta v_j^k = v_j^k + \left(-\eta \frac{\partial E(x,p)^k}{\partial v_j^k}\right) \tag{3.9}$$

The derivatives of the error function with respect to w_j and v_j may be obtained as

$$\frac{\partial E(x,p)}{\partial w_j} = \frac{\partial}{\partial w_j}\left(\sum_{i=1}^{h}\frac{1}{2}\left\{\frac{d^n y_t(x_i,p)}{dx^n} - f\left(x_i, y_t(x_i), \frac{dy_t(x_i,p)}{dx}, \ldots, \frac{d^{n-1}y_t(x_i,p)}{dx^{n-1}}\right)\right\}^2\right) \quad (3.10)$$

$$\frac{\partial E(x,p)}{\partial v_j} = \frac{\partial}{\partial v_j}\left(\sum_{i=1}^{h}\frac{1}{2}\left\{\frac{d^n y_t(x_i,p)}{dx^n} - f\left(x_i, y_t(x_i), \frac{dy_t(x_i,p)}{dx}, \ldots, \frac{d^{n-1}y_t(x_i,p)}{dx^{n-1}}\right)\right\}^2\right) \quad (3.11)$$

For a clear understanding, we include next the formulations of first- and second-order IVPs.

3.2.2.1 Formulation of First-Order IVPs

Let us consider a first-order ODE as follows:

$$\frac{dy}{dx} = f(x,y) \quad x \in [a,b] \quad (3.12)$$

with initial condition $y(a) = A$.

In this case, the ANN trial solution is written as

$$y_t(x,p) = A + (x-a)N(x,p) \quad (3.13)$$

where $N(x,p)$ is the output of feed-forward network with input data $x = (x_1, x_2, \ldots, x_h)^T$ and parameters p.

Differentiating Equation 3.13 with respect to x, we have

$$\frac{dy_t(x,p)}{dx} = (x-a)N(x,p) + \frac{dN}{dx} \quad (3.14)$$

The error function for this case may be formulated as

$$E(x,p) = \sum_{i=1}^{h}\frac{1}{2}\left(\frac{dy_t(x_i,p)}{dx} - f\left(x_i, y_t(x_i,p)\right)\right)^2 \quad (3.15)$$

3.2.2.2 Formulation of Second-Order IVPs

A second-order ODE may be written in general as

$$\frac{d^2y}{dx^2} = f\left(x, y, \frac{dy}{dx}\right) \quad x \in [a,b] \quad (3.16)$$

subject to $y(a) = A$, $y'(a) = A'$.

The ANN trial solution is expressed as

$$y_t(x,p) = A + A'(x-a) + (x-a)^2 N(x,p) \tag{3.17}$$

where $N(x,p)$ is the output of the ANN with input x and parameters p. The trial solution $y_t(x,p)$ satisfies the initial conditions.

From Equation 3.17, we have (by differentiating)

$$\frac{dy_t(x,p)}{dx} = A' + 2(x-a)N(x,p) + (x-a)^2 \frac{dN}{dx} \tag{3.18}$$

and

$$\frac{d^2y_t(x,p)}{dx^2} = 2N(x,p) + 4(x-a)\frac{dN}{dx} + (x-a)^2 \frac{d^2N}{dx^2} \tag{3.19}$$

The error function to be minimized for the second-order ODE is found to be

$$E(x,p) = \sum_{i=1}^{h} \frac{1}{2}\left(\frac{d^2y_t(x_i,p)}{dx^2} - f\left(x_i, y_t(x_i,p), \frac{dy_t(x_i,p)}{dx} \right) \right)^2 \tag{3.20}$$

As discussed in Section 3.2.2, the weights from the input layer to the hidden layer and from the hidden layer to the output layer are modified according to the unsupervised back-propagation learning algorithm.

The derivatives of the error function with respect to w_j and v_j are written as

$$\frac{\partial E(x,p)}{\partial w_j} = \frac{\partial}{\partial w_j}\left(\sum_{i=1}^{h} \frac{1}{2}\left(\frac{d^2y_t(x_i,p)}{dx^2} - f\left[x_i, y_t(x_i,p), \frac{dy_t(x_i,p)}{dx} \right] \right)^2 \right) \tag{3.21}$$

$$\frac{\partial E(x,p)}{\partial v_j} = \frac{\partial}{\partial v_j}\left(\sum_{i=1}^{h} \frac{1}{2}\left(\frac{d^2y_t(x_i,p)}{dx^2} - f\left[x_i, y_t(x_i,p), \frac{dy_t(x_i,p)}{dx} \right] \right)^2 \right) \tag{3.22}$$

3.2.3 Formulation of BVPs

Next, we include the formulation of second- and fourth-order BVPs.

3.2.3.1 Formulation of Second-Order BVPs

Let us consider a second-order BVP [12]

$$\frac{d^2y}{dx^2} = f\left(x, y, \frac{dy}{dx} \right) \quad x \in [a,b] \tag{3.23}$$

subject to boundary conditions $y(a) = A$, $y(b) = B$.

The corresponding ANN trial solution for this BVP is formulated as

$$y_t(x,p) = \frac{bA - aB}{b-a} + \frac{B-A}{b-a}x + (x-a)(x-b)N(x,p) \tag{3.24}$$

Differentiating Equation 3.24, we have

$$\frac{dy_t(x,p)}{dx} = \frac{B-A}{b-a} + (x-b)N(x,p) + (x-a)N(x,p) + (x-a)(x-b)\frac{dN}{dx} \tag{3.25}$$

As such, the error function may be obtained as

$$E(x,p) = \sum_{i=1}^{h} \frac{1}{2}\left(\frac{d^2 y_t(x_i,p)}{dx^2} - f\left(x_i, y_t(x_i,p), \frac{dy_t(x_i,p)}{dx} \right) \right)^2 \tag{3.26}$$

3.2.3.2 Formulation of Fourth-Order BVPs

A general fourth-order differential equation is considered as [12]

$$\frac{d^4 y}{dx^4} = f\left(x, y, \frac{dy}{dx}, \frac{d^2 y}{dx^2}, \frac{d^3 y}{dx^3} \right) \tag{3.27}$$

with boundary conditions

$$y(a) = A, \quad y(b) = B, \quad y'(a) = A', \quad y'(b) = B'.$$

The ANN trial solution for the previous fourth-order differential equation satisfying the boundary conditions is constructed as

$$y_t(x,p) = Z(x) + M(x)N(x,p) \tag{3.28}$$

The trial solution satisfies the following relations:

$$\left.\begin{aligned} Z(a) &= A, \\ Z(b) &= B, \\ Z'(a) &= A', \\ Z'(b) &= B'. \end{aligned}\right\} \tag{3.29}$$

$$\left.\begin{aligned} &M(a)N(a,p) = 0, \\ &M(b)N(b,p) = 0, \\ &M(a)N'(a,p) + M'(a)N(a,p) = 0, \\ &M(b)N'(b,p) + M'(b)N(b,p) = 0. \end{aligned}\right\} \tag{3.30}$$

The function $M(x)$ is chosen as $M(x)=(x-a)^2(x-b)^2$, which satisfies the set of equations in (3.30). Here, $Z(x)=a'x^4+b'x^3+c'x^2+d'x$ is the general polynomial of degree four, where a', b', c', and d' are constants. From the set of Equations 3.29, we have

$$\left.\begin{aligned} a'a^4+b'a^3+c'a^2+d'a &= A \\ a'b^4+b'b^3+c'b^2+d'b &= B \\ 4a'a^3+3b'a^2+2c'a+d' &= A' \\ 4a'b^3+3b'b^2+2c'b+d' &= B' \end{aligned}\right\} \tag{3.31}$$

Solving this system of four equations with four unknowns, we obtain the general form of the polynomial $Z(x)$.

Here, the error function is expressed as

$$E(x,p)=\sum_{i=1}^{h}\frac{1}{2}\left(\frac{d^4y_t(x_i,p)}{dx^4}-f\left[x_i,y_t(x_i,p),\frac{d^2y_t(x_i,p)}{dx^2},\frac{d^3y_t(x_i,p)}{dx^3}\right]\right)^2 \tag{3.32}$$

3.2.4 Formulation of a System of First-Order ODEs

We consider now the following system of first-order ODEs [12]:

$$\frac{dy_r}{dx}=f_r\left(x,y_1,\ldots,y_\ell\right) \quad r=1,2,\ldots,\ell \text{ and } x\in\left[a,b\right] \tag{3.33}$$

subject to $y_r(a)=A_r$, $r=1,2,\ldots,\ell$.

The corresponding ANN trial solution has the following form:

$$y_{t_r}(x,p_r)=A_r+(x-a)N_r(x,p_r) \quad \forall r=1,2,\ldots,\ell \tag{3.34}$$

For each r, $N_r(x,p_r)$ is the output of the multilayer ANN with input x and parameter p_r.

From Equation 3.34, we have

$$\frac{dy_{t_r}(x,p_r)}{dx}=(x-a)N_r(x,p_r)+\frac{dN_r}{dx} \quad \forall r=1,2,\ldots,\ell \tag{3.35}$$

Then, the corresponding error function with adjustable network parameters may be written as

$$E(x,p)=\sum_{i=1}^{h}\sum_{r=1}^{\ell}\frac{1}{2}\left[\frac{dy_{t_r}(x_i,p_r)}{dx}-f_r\left(x_i,y_{t_1}(x_i,p_1),\ldots,y_{t_\ell}(x_i,p_\ell)\right)\right]^2 \tag{3.36}$$

For a system of first-order ODEs (Equation 3.36), we have the derivatives of the error function with respect to w_j and v_j as follows:

$$\frac{\partial E(x,p)}{\partial w_j} = \frac{\partial}{\partial w_j}\left(\sum_{i=1}^{h}\frac{1}{2}\left[\begin{array}{l}\left\{\dfrac{dy_{t_1}(x_i,p_1)}{dx} - f_1\left(x_i,y_{t_1}(x_i,p_1),\ldots,y_{t_\ell}(x_i,p_\ell)\right)\right\}\\ +\left\{\dfrac{dy_{t_2}(x_i,p_2)}{dx} - f_2\left(x_i,y_{t_1}(x_i,p_1),\ldots,y_{t_\ell}(x_i,p_\ell)\right)\right\}\\ +\cdots+\left\{\dfrac{dy_{t_\ell}\left(x_i,p_\ell\right)}{dx} - f_l\left(x_i,y_{t_1}(x_i,p_1),\ldots,y_{t_\ell}(x_i,p_\ell)\right)\right\}\end{array}\right]\right)^2 \tag{3.37}$$

$$\frac{\partial E(x,p)}{\partial v_j} = \frac{\partial}{\partial v_j}\left(\sum_{i=1}^{h}\frac{1}{2}\left[\begin{array}{l}\left\{\dfrac{dy_{t_1}(x_i,p_1)}{dx} - f_1\left(x_i,y_{t_1}(x_i,p_1),\ldots,y_{t_\ell}(x_i,p_\ell)\right)\right\}\\ +\left\{\dfrac{dy_{t_2}(x_i,p_2)}{dx} - f_2\left(x_i,y_{t_1}(x_i,p_1),\ldots,y_{t_\ell}(x_i,p_\ell)\right)\right\}\\ +\cdots+\left\{\dfrac{dy_{t_\ell}(x_i,p_\ell)}{dx} - f_l\left(x_i,y_{t_1}(x_i,p_1),\ldots,y_{t_\ell}(x_i,p_\ell)\right)\right\}\end{array}\right]\right)^2 \tag{3.38}$$

It may be noted that the detailed procedure of handling IVPs and BVPs using a single-layer ANN model is discussed in subsequent chapters.

Next, we address the computation of the gradient of ODEs using the traditional multilayer ANN model.

3.2.5 Computation of Gradient of ODEs for Multilayer ANN Model

Error computation not only involves the output but also the derivative of the network output with respect to its input [6]. So it requires finding the gradient of the network derivative with respect to its input. Let us now consider a multilayer ANN with one input node x that has h number of data $(x=(x_1,x_2,\ldots,x_h)^T)$, a hidden layer with m nodes, and one output unit. The network output $N(x,p)$ is formulated as

$$N(x,p) = \sum_{j=1}^{m} v_j s(z_j) \tag{3.39}$$

The derivative of $N(x,p)$ with respect to input x is

$$\frac{d^kN}{dx^k} = \sum_{j=1}^{m} v_j w_j^k s_j^{(k)} \tag{3.40}$$

where $s = s(z_j)$ and $s^{(k)}$ denotes the kth-order derivative of an activation function.

The gradient of the output with respect to the network parameters of the ANN may be formulated as

$$\frac{\partial N}{\partial v_j} = s(z_j) \tag{3.41}$$

$$\frac{\partial N}{\partial u_j} = v_j s'(z_j) \tag{3.42}$$

$$\frac{\partial N}{\partial w_j} = v_j s'(z_j) x \tag{3.43}$$

N_α is the derivative of the network output with respect to any of its input and

$$N_\alpha = D^n N = \sum_{j=1}^{m} v_j P_j s_j^{(\Lambda)} \tag{3.44}$$

We have the relation

$$P_j = \prod_{k=1,2,\ldots,n} w_j^k \tag{3.45}$$

Derivatives of N_α with respect to other parameters are given as

$$\frac{\partial N_\alpha}{\partial v_j} = P_j s_j^{(\Lambda)} \tag{3.46}$$

$$\frac{\partial N_\alpha}{\partial u_j} = v_j P_j s_j^{(\Lambda+1)} \tag{3.47}$$

$$\frac{\partial N_\alpha}{\partial w_j} = x v_j P_j s_j^{(\Lambda+1)} + v_j \Lambda_j w_j^{\Lambda_i - 1}, \quad i = 1, 2, \ldots, n \tag{3.48}$$

where Λ denotes the order of the derivative.

After getting all the derivatives, we can find the gradient of error. Using the error back-propagation learning method for unsupervised training, we may minimize the error function as per the desired accuracy.

3.3 First-Order Linear ODEs

Two first-order linear ODEs are given in Examples 3.1 and 3.2.

Example 3.1

A first-order ODE is

$$\frac{dy(x)}{dx} = 2x + 1 \quad x \in [0,1]$$

subject to $y(0) = 0$.

According to Section 3.2.2.1 (Equation 3.13), the trial solution may be written as

$$y_t(x, p) = xN(x, p).$$

The network is trained for six equidistant points in [0, 1] and five sigmoid hidden nodes. Table 3.1 shows the neural results at different error values and the convergence of the neural output up to the given accuracy. The weights are selected randomly. Analytical and ANN results with an accuracy of 0.0001 are plotted in Figure 3.2. The error (between analytical and ANN results) is plotted in Figure 3.3. Neural results for some testing points with

TABLE 3.1

Neural Results for Different Errors (Example 3.1)

	Error →	**0.5**	**0.1**	**0.05**	**0.01**	**0.005**	**0.001**	**0.0005**	**0.0001**
Input Data	**Exact ↓**				**Neural Results**				
0	0	0	0	0	0	0	0	0	0
0.2	0.2400	0.2098	0.2250	0.2283	0.2312	0.2381	0.2401	0.2407	0.2418
0.4	0.5600	0.4856	0.4778	0.4836	0.4971	0.5395	0.5410	0.5487	0.5503
0.6	0.9600	0.9818	0.7889	0.7952	0.8102	0.8672	0.9135	0.9418	0.9562
0.8	1.4400	1.7915	1.1390	1.1846	1.2308	1.2700	1.3341	1.3722	1.4092
1.0	2.0000	2.8339	1.4397	1.5401	1.6700	1.7431	1.8019	1.8157	1.9601

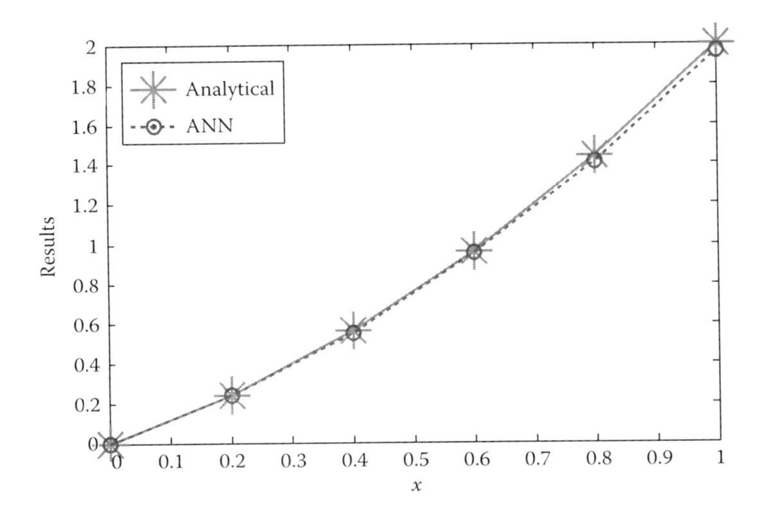

FIGURE 3.2
Plot of analytical and ANN results (Example 3.1).

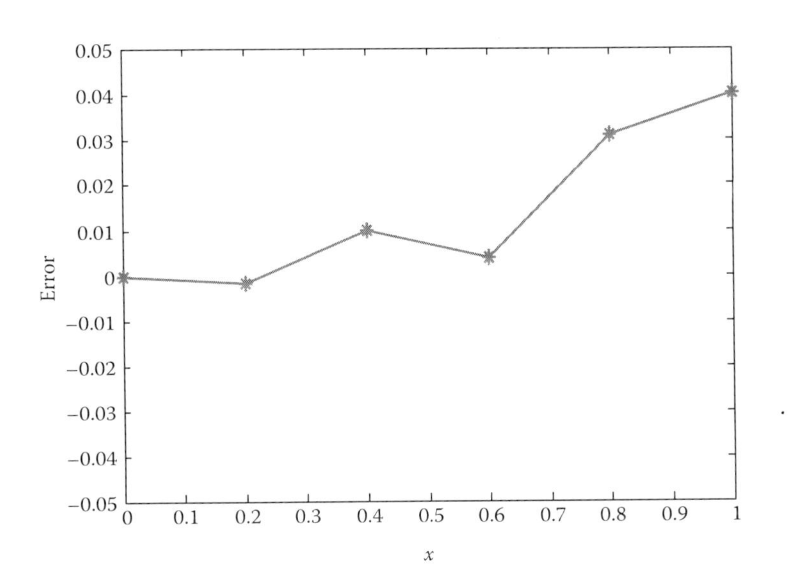

FIGURE 3.3
Error plot between analytical and ANN results (Example 3.1).

TABLE 3.2

Analytical and Neural Results for Testing Points (Inside the Domain) (Example 3.1)

Testing Points	0.8235	0.6787	0.1712	0.3922	0.0318	0.9502
Exact	1.5017	1.1393	0.2005	0.5460	0.0328	1.8531
ANN	1.6035	1.3385	0.2388	0.5701	0.0336	1.9526

TABLE 3.3

Analytical and Neural Results for Testing Points (Outside the Domain) (Example 3.1)

Testing Points	1.2769	1.1576	1.0357	1.3922	1.4218	1.2147
Exact	2.9074	2.4976	2.1084	3.3304	3.4433	2.6902
ANN	2.8431	2.4507	2.0735	3.4130	3.5163	2.6473

an accuracy of 0.0001 are shown in Tables 3.2 (inside the domain) and 3.3 (outside the domain).

Example 3.2

In this example, we have considered a first-order linear ODE

$$\frac{dy}{dx} + 0.2y = e^{-0.2x}\cos x \quad x \in [0,1]$$

subject to $y(0) = 0$.

The trial solution is formulated as

$$y_t(x,p) = xN(x,p).$$

We have trained the network for 10 equidistant points in [0, 1] and 4 and 5 sigmoid hidden nodes. Table 3.4 shows the ANN results at different error values for four hidden nodes. ANN results with five hidden nodes at different error values have been given in Table 3.5. A comparison between analytical and ANN results for four and five hidden nodes (at error 0.001) is presented in Figures 3.4 and 3.5, respectively.

TABLE 3.4

Comparison among Analytical and ANN Results at Different Error Values for Four Hidden Nodes (Example 3.2)

Input Data	Analytical	ANN Results at Error = 0.1	ANN Results at Error = 0.01	ANN Results at Error = 0.001
0	0	0	0	0
0.1	0.0979	0.0879	0.0938	0.0967
0.2	0.1909	0.1755	0.1869	0.1897
0.3	0.2783	0.2646	0.2795	0.2796
0.4	0.3595	0.3545	0.3580	0.3608
0.5	0.4338	0.4445	0.4410	0.4398
0.6	0.5008	0.5325	0.5204	0.5208
0.7	0.5601	0.6200	0.6085	0.5913
0.8	0.6113	0.7023	0.6905	0.6695
0.9	0.6543	0.7590	0.7251	0.6567
1	0.6889	0.8235	0.7936	0.6825

TABLE 3.5

Comparison among Analytical and ANN Results at Different Error Values for Five Hidden Nodes (Example 3.2)

Input Data	Analytical	ANN Results at Error = 0.1	ANN Results at Error = 0.01	ANN Results at Error = 0.001
0	0	0	0	0
0.1	0.0979	0.0900	0.0949	0.0978
0.2	0.1909	0.1805	0.1874	0.1901
0.3	0.2783	0.2714	0.2754	0.2788
0.4	0.3595	0.3723	0.3605	0.3600
0.5	0.4338	0.4522	0.4469	0.4389
0.6	0.5008	0.5395	0.5213	0.5166
0.7	0.5601	0.6198	0.6077	0.5647
0.8	0.6113	0.6995	0.6628	0.6111
0.9	0.6543	0.7487	0.7021	0.6765
1	0.6889	0.8291	0.7790	0.7210

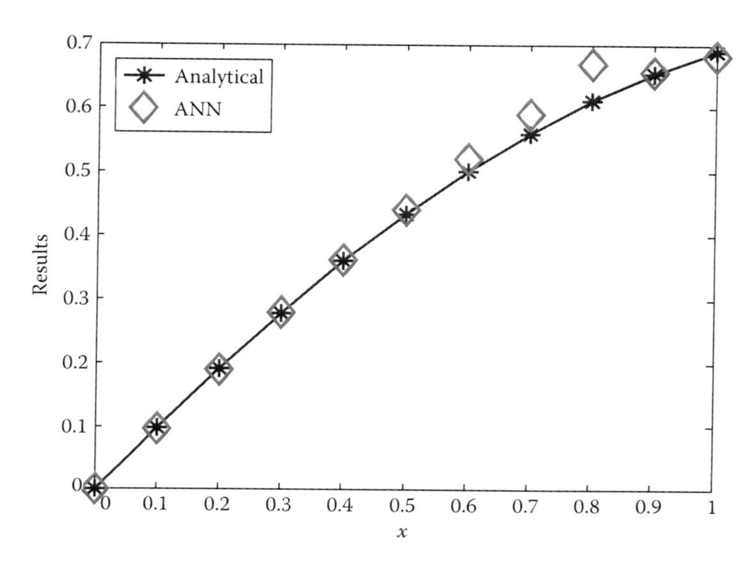

FIGURE 3.4
Plot of analytical and ANN results at error 0.001 for four hidden nodes (Example 3.2).

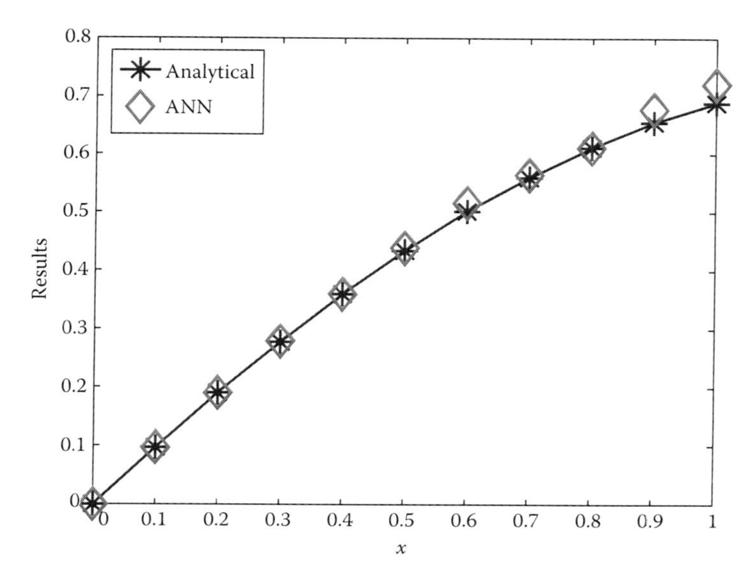

FIGURE 3.5
Plot of analytical and ANN results at error 0.001 for five hidden nodes (Example 3.2).

3.4 Higher-Order ODEs

Example 3.3

In this example, a second-order differential equation with initial conditions that describes a model of an undamped free vibration spring mass system is considered.

$$\frac{d^2y}{dx^2} + y = 0 \quad x \in [0,1]$$

with initial conditions $y(0)=0$ and $y'(0)=1$.

As discussed in Section 3.2.2.2 (Equation 3.17), the ANN trial solution is written as

$$y_t(x,p) = x + x^2 N(x,p)$$

The network has been trained here for 10 equidistant points in [0, 1] and 7 hidden nodes. We have considered the sigmoid function as the activation function. A comparison between analytical and neural approximate results with random initial weights is given in Table 3.6. These analytical and neural results for random initial weights are also depicted in Figure 3.6. Finally, the error plot between analytical and ANN results is presented in Figure 3.7.

TABLE 3.6

Analytical and ANN Results (Example 3.3)

Input Data	Analytical	ANN
0	0	0
0.1	0.0998	0.0996
0.2	0.1987	0.1968
0.3	0.2955	0.2905
0.4	0.3894	0.3808
0.5	0.4794	0.4714
0.6	0.5646	0.5587
0.7	0.6442	0.6373
0.8	0.7174	0.7250
0.9	0.7833	0.8043
1	0.8415	0.8700

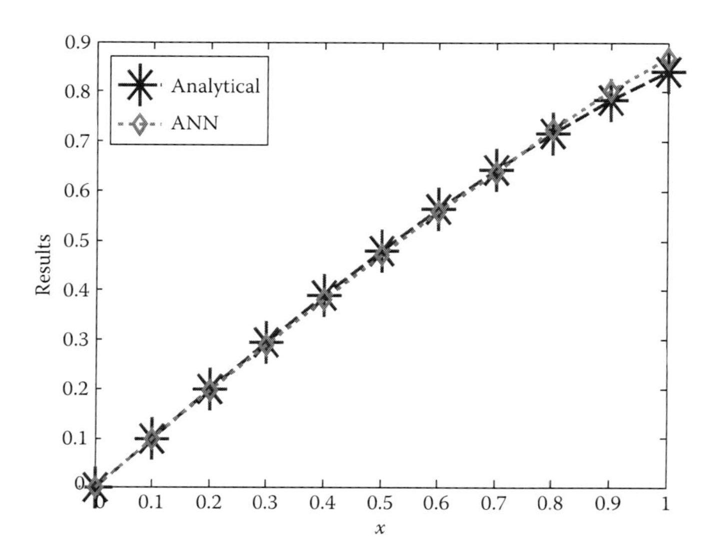

FIGURE 3.6
Plot of analytical and ANN results (Example 3.3).

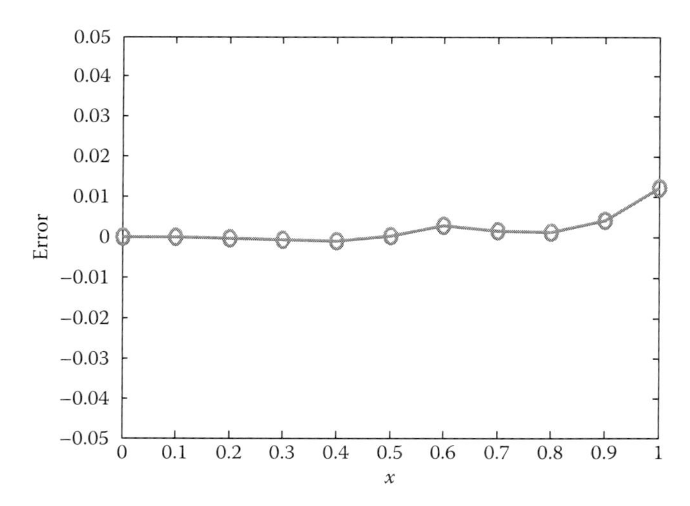

FIGURE 3.7
Error plot between analytical and ANN results (Example 3.3).

3.5 System of ODEs

Example 3.4

In this example, a system of coupled first-order ODEs has been considered [6]

$$\left.\begin{aligned}\frac{dy_1}{dx} &= \cos(x) + y_1^2 + y_2 - \left(1 + x^2 + \sin^2(x)\right)\\ \frac{dy_2}{dx} &= 2x - (1 + x^2)\sin(x) + y_1 y_2\end{aligned}\right\}\quad x \in [0,2]$$

with initial conditions $y_1(0) = 0$ and $y_2(0) = 1$.

The ANN trial solutions are (discussed in Section 3.2.4, Equation 3.34) as follows:

$$\left.\begin{aligned}y_{t_1}(x) &= xN_1(x, p_1)\\ y_{t_2}(x) &= 1 + xN_2(x, p_2)\end{aligned}\right\}$$

Twenty equidistant points in [0, 2] and five weights from the input layer to the hidden layer and from the hidden layer to the output layer are considered. A comparison between analytical $(y_1(x), y_2(x))$ and ANN results $(y_{t1}(x), y_{t2}(x))$ is given in Table 3.7. The error plot between analytical and ANN results is depicted in Figure 3.8.

TABLE 3.7

Comparison between Analytical and ANN Results (Example 3.4)

Input Data	Analytical [6] $y_1(x)$	ANN $y_{t_1}(x)$	Analytical [6] $y_2(x)$	ANN $y_{t_2}(x)$
0	0	0.0001	1.0000	1.0000
0.1000	0.0998	0.1019	1.0100	1.0030
0.2000	0.1987	0.2027	1.0400	1.0460
0.3000	0.2955	0.2998	1.0900	1.0973
0.4000	0.3894	0.3908	1.1600	1.1624
0.5000	0.4794	0.4814	1.2500	1.2513
0.6000	0.5646	0.5689	1.3600	1.3628
0.7000	0.6442	0.6486	1.4900	1.4921
0.8000	0.7174	0.7191	1.6400	1.6425
0.9000	0.7833	0.7864	1.8100	1.8056
1.0000	0.8415	0.8312	2.0000	2.0046
1.1000	0.8912	0.8897	2.2100	2.2117
1.2000	0.9320	0.9329	2.4400	2.4383
1.3000	0.9636	0.9642	2.6900	2.6969
1.4000	0.9854	0.9896	2.9600	2.9640
1.5000	0.9975	0.9949	3.2500	3.2442
1.6000	0.9996	0.9960	3.5600	3.5679
1.7000	0.9917	0.9907	3.8900	3.8970
1.8000	0.9738	0.9810	4.2400	4.2468
1.9000	0.9463	0.9470	4.6100	4.6209
2.0000	0.9093	0.9110	5.0000	5.0012

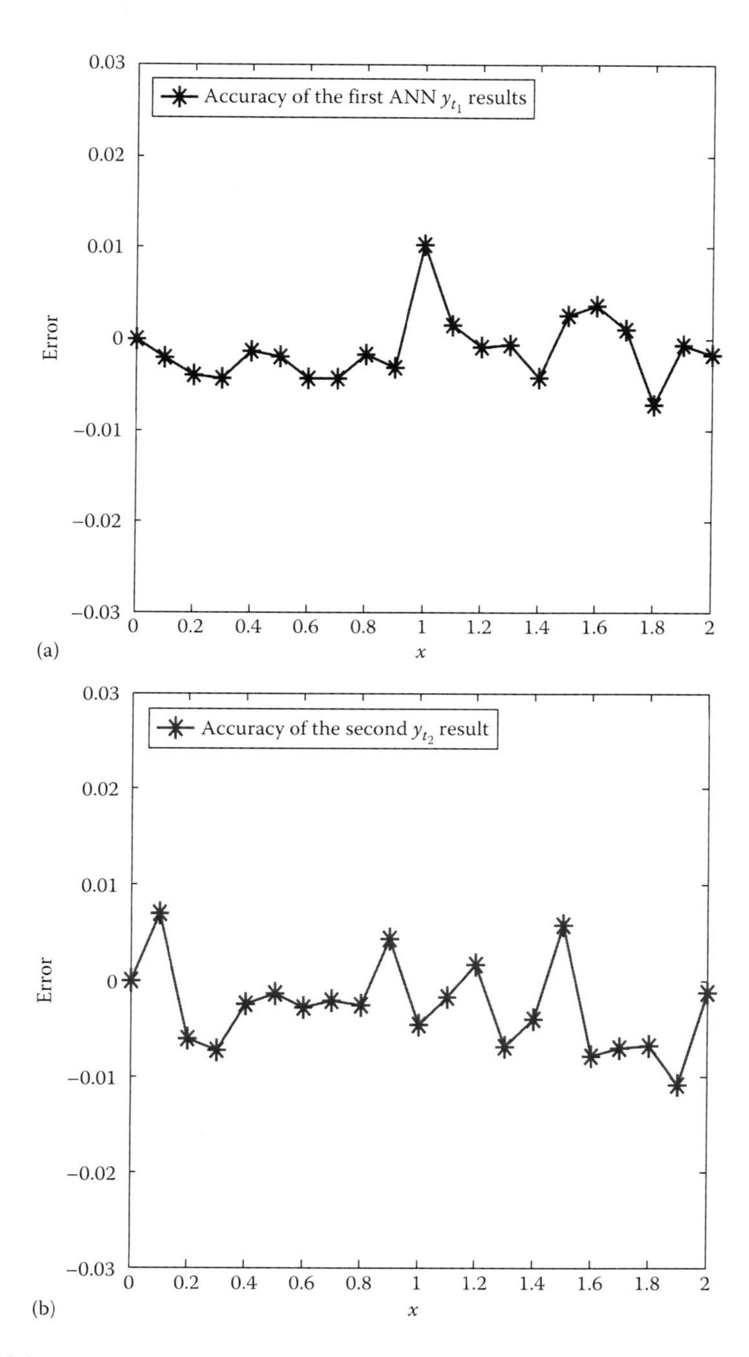

FIGURE 3.8
Error plots between analytical and ANN results (Example 3.4). (a) Error plot of y_{t_1}. (b) Error plot of y_{t_2}.

References

1. J.M. Zurada. *Introduction to Artificial Neural Systems*. West Publishing Co., St. Paul, MN, 1992.
2. S.S. Haykin. *Neural Networks: A Comprehensive Foundation*. Prentice Hall Inc., Upper Saddle River, NJ, 1999.
3. H. Lee and I.S. Kang. Neural algorithm for solving differential equations. *Journal of Computational Physics*, 91(1): 110–131, November 1990.
4. A.J. Meade and A.A. Fernandez. The numerical solution of linear ordinary differential equations by feed forward neural networks. *Mathematical and Computer Modelling*, 19(2): 1–25, June 1994.
5. A.J. Meade and A.A. Fernandez. Solution of nonlinear ordinary differential equations by feed forward neural networks. *Mathematical and Computer Modelling*, 20(9): 19–44, November 1994.
6. E. Lagaris, A. Likas, and D.I. Fotiadis. Artificial neural networks for solving ordinary and partial differential equations. *IEEE Transactions on Neural Networks*, 9(5): 987–1000, September 1998.
7. M. Kumar and N. Yadav. Multilayer perceptrons and radial basis function neural network methods for the solution of differential equations a survey. *Computers and Mathematics with Applications*, 62(10): 3796–3811, November 2011.
8. K.I. Ibraheem and B.M. Khalaf. Shooting neural networks algorithm for solving boundary value problems in ODEs. *Applications and Applied Mathematics*, 6(11): 1927–1941, June 2011.
9. B.-A. Liu and B. Jammes. Solving ordinary differential equations by neural networks. *Proceeding of 13th European Simulation Multi-Conference Modelling and Simulation: A Tool for the Next Millennium*, Warsaw, Poland, June 1999.
10. S. Mall and S. Chakraverty. Comparison of artificial neural network architecture in solving ordinary differential equations. *Advances in Artificial Neural Systems*, 2013: 1–24, October 2013.
11. S. Chakraverty and S. Mall. Regression based weight generation algorithm in neural network for solution of initial and boundary value problems. *Neural Computing and Applications*, 25(3): 585–594, September 2014.
12. A. Malek and R.S. Beidokhti. Numerical solution for high order deferential equations, using a hybrid neural network—Optimization method. *Applied Mathematics and Computation*, 183(1): 260–271, December 2006.
13. H.S. Yazdi, M. Pakdaman, and H. Modaghegh. Unsupervised kernel least mean square algorithm for solving ordinary differential equations. *Nerocomputing*, 74(12–13): 2062–2071, June 2011.
14. G. Tsoulos and I.E. Lagaris. Solving differential equations with genetic programming. *Genetic Programming and Evolvable Machines*, 7(1): 33–54, March 2006.
15. D.R. Parisi, M.C. Mariani, and M.A. Laborde. Solving differential equations with unsupervised neural networks. *Chemical Engineering and Processing* 42: 715–721, September 2003.

4

Regression-Based ANN

In this chapter, the regression-based neural network (RBNN) model has been used for solving first- and higher-order ordinary differential equations (ODEs) [1–4]. The trial solution of the differential equation has been obtained by using the RBNN model for single input and single output (SISO) system. The number of nodes in the hidden layer has been fixed according to the degree of the polynomial in regression fitting, and the coefficients involved are taken as initial weights to start with neural training. Here, the unsupervised error back-propagation method has been used for minimizing the error function. Modifications of the parameters are done without the use of any optimization technique. Initial weights from input to hidden and from hidden to output layer are taken as combination of random (arbitrary) as well as by RBNN. In this chapter, a variety of initial and boundary value problems have been solved and the results with arbitrary and regression-based initial weights are compared.

It has been already pointed out in the previous paragraph that the RBNN model may be used to fix the number of nodes in the hidden layer using regression analysis.

As such, Chakraverty and his coauthors [5,6] have investigated various application problems using RBNN model. Prediction of the response of structural systems subject to earthquake motions has been investigated by Chakraverty et al. [5] using the RBNN model. Chakraverty et al. [6] studied vibration frequencies of annular plates using RBNN. Recently, Mall and Chakraverty [1–4] proposed the RBNN model for solving initial/boundary value problems (BVPs) of ODEs.

4.1 Algorithm of RBNN Model

RBNN model has been investigated by Chakraverty et al. [5,6] for various application problems. Let us consider training patterns as $\{(x_1,y_1),(x_2,y_2),\ldots,(x_n,y_n)\}$. For every value of x_i, we may find y_i crudely by other traditional numerical methods. But these methods (traditional) are usually iterative in nature, where we fix the step size before the start of the computation. After the solution is obtained, if we want to know the solution in between steps, then again we have to iterate the procedure from the initial stage. ANN may be one of the ways where we may overcome this repetition of iterations. Also, ANN has an inherent advantage over numerical methods [7,8].

As mentioned earlier, the initial weights from the input layer to the hidden layer are generated by coefficients of regression analysis. Let x and y be the input and output patterns, then a polynomial of degree four is written as [9,10]

$$p(x) = a_0 + a_1x + a_2x^2 + a_3x^3 + a_4x^4 \tag{4.1}$$

where a_0, a_1, a_2, a_3, and a_4 are coefficients of this polynomial, which may be obtained by using the least-square fit method. These constants may be now taken as the initial weights from the input layer to the hidden layer. Then, we calculate the output of the nodes of the hidden layer by using the activation functions [3,5,6]

$$h_0^i = \frac{1}{1+e^{-\lambda a_0}} \quad i = 1,2,3,\ldots,n \tag{4.2}$$

$$h_1^i = \frac{1}{1+e^{-\lambda a_1 x_i}} \quad i = 1,2,3,\ldots,n \tag{4.3}$$

$$h_2^i = \frac{1}{1+e^{-\lambda a_2 x_i^2}} \quad i = 1,2,3,\ldots,n \tag{4.4}$$

$$h_3^i = \frac{1}{1+e^{-\lambda a_3 x_i^3}} \quad i = 1,2,3,\ldots,n \tag{4.5}$$

$$h_4^i = \frac{1}{1+e^{-\lambda a_4 x_i^4}} \quad i = 1,2,3,\ldots,n \tag{4.6}$$

The regression analysis is applied again to find the output of the network by the relation

$$c_0h_0^i + c_1h_1^i + c_2h_2^i + c_3h_3^i + c_4h_4^i = y, \quad i = 1,2,3,\ldots,n \tag{4.7}$$

where c_0, c_1, c_2, c_3, and c_4 are coefficients of this multivariate linear regression polynomial and may be obtained by the least-square fit method. Subsequently, these constants are then considered as the initial weights from the hidden layer to the output layer.

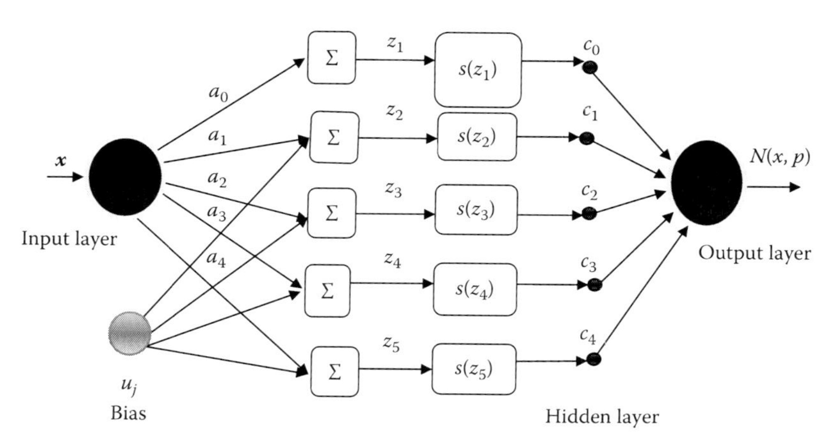

FIGURE 4.1
RBNN architecture with single input node and single output node.

4.2 Structure of RBNN Model

A three-layer RBNN model has been considered for the present problem. Figure 4.1 shows the neural network architecture, in which the input layer consists of a single input unit along with biases (u_j), a hidden layer (contains five hidden nodes), and an output layer consisting of one output node. The number of nodes in the hidden layer depends upon the degree of regression fitting that is proposed here. If an nth-degree polynomial is considered, then the number of nodes in the hidden layer will be $n+1$ and coefficients (constants, say, a_i, c_i) of the polynomial are considered as the initial weights from the input layer to the hidden layer as well as from the hidden layer to the output layer. Architecture of the network with a fourth-degree polynomial is shown in Figure 4.1.

4.3 Formulation and Learning Algorithm of RBNN Model

The RBNN trial solution $y_t(x,p)$ for ODEs with network parameters p (weights, biases) may be written in the form

$$y_t(x,p) = A(x) + F(x, N(x,p)) \tag{4.8}$$

The first term $A(x)$ on the right-hand side does not contain adjustable parameters and satisfies only initial/boundary conditions, whereas the second term $F(x,N(x,p))$ contains the single output $N(x,p)$ of RBNN with input x and adjustable parameters p.

Here, we consider a three-layer network with one input node, one hidden layer consisting of m number of nodes, and one output $N(x,p)$. For every input x and parameters p, the output is defined as

$$N(x,p) = \sum_{j=1}^{m} v_j s(z_j) \tag{4.9}$$

where $z_j = w_j x + u_j$, w_j denotes the weight from the input unit to the hidden unit j, v_j denotes the weight from the hidden unit j to the output unit, u_j are the biases, and $s(z_j)$ is the activation function (sigmoid, tangent hyperbolic, etc.).

In this regard, the formulation of first- and second-order initial value problems (IVPs) has been discussed in Sections 3.2.2.1 and 3.2.2.2 (Equations 3.13 and 3.17), respectively. Training the neural network means updating the parameters (weights and biases) so that the error values converge to the desired accuracy. The unsupervised error back-propagation learning algorithm (Equations 3.8 through 3.11) has been used to update the network parameters (weights and biases) from the input layer to the hidden layer and from the hidden layer to the output layer and to minimize the error function of the RBNN model.

4.4 Computation of Gradient for RBNN Model

Error computation not only involves the output but also the derivatives of the network output with respect to its input and parameters. So it requires finding out the gradient of the network derivatives with respect to their inputs. For minimizing the error function $E(x,p)$, that is, to update the network parameters (weights and biases), we differentiate $E(x,p)$ with respect to the parameters. The gradient of the network output with respect to its inputs is computed in Section 3.2.5.

4.5 First-Order Linear ODEs

Here, we have taken three first-order ODEs to show the reliability of the RBNN model. Also, the accuracy of results of the proposed RBNN model has been shown in the tables and figures.

Example 4.1

Let us consider a first-order ODE

$$\frac{dy}{dx} = x + y \quad x \in [0,1]$$

with initial condition $y(0) = 1$.

The RBNN trial solution in this case is written as

$$y_t(x,p) = 1 + xN(x,p).$$

The network has been trained for 20 equidistant points in [0, 1] and four hidden nodes are fixed according to regression analysis with a third-degree polynomial for the RBNN model. Six hidden nodes have been considered for the traditional ANN model. Here, the activation function is the sigmoid function. We have compared analytical results with neural approximate results with random and regression-based weights in Table 4.1. One may very well

TABLE 4.1

Analytical and Neural Results with Arbitrary (Random) and Regression-Based Weights (Example 4.1)

Input Data	Analytical	ANN Results with Random Weights	RBNN
0	1.0000	1.0000	1.0000
0.0500	1.0525	1.0533	1.0522
0.1000	1.1103	1.1092	1.1160
0.1500	1.1737	1.1852	1.1732
0.2000	1.2428	1.2652	1.2486
0.2500	1.3181	1.3320	1.3120
0.3000	1.3997	1.4020	1.3975
0.3500	1.4881	1.5007	1.4907
0.4000	1.5836	1.5771	1.5779
0.4500	1.6866	1.6603	1.6631
0.5000	1.7974	1.8324	1.8006
0.5500	1.9165	1.8933	1.9132
0.6000	2.0442	2.0119	2.0615
0.6500	2.1811	2.1380	2.1940
0.7000	2.3275	2.3835	2.3195
0.7500	2.4840	2.4781	2.4825
0.8000	2.6511	2.6670	2.6535
0.8500	2.8293	2.8504	2.8305
0.9000	3.0192	3.006	3.0219
0.9500	3.2214	3.2482	3.2240
1.0000	3.4366	3.4281	3.4402

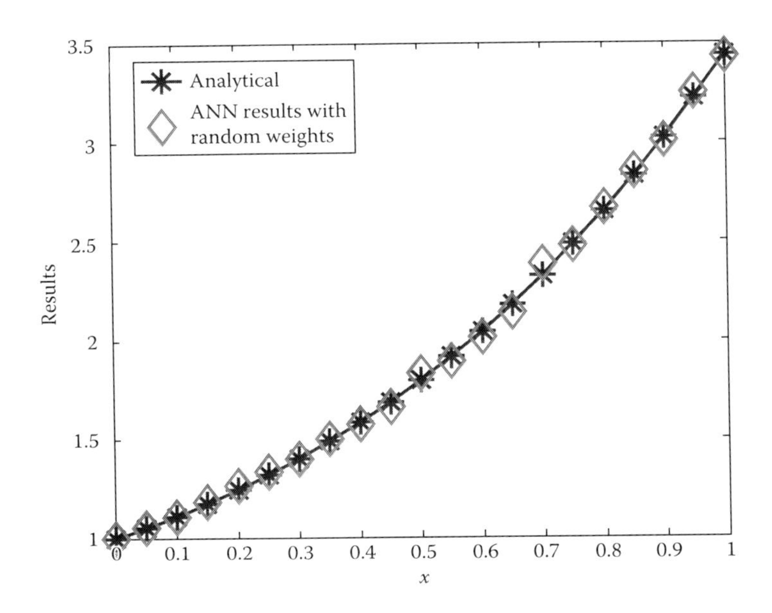

FIGURE 4.2
Plot of analytical and neural results with arbitrary weights (Example 4.1).

observe that the better results are got by using the RBNN method, which is tabulated in the third column. Figure 4.2 shows a comparison between analytical and neural results when initial weights are random. Analytical and neural results for regression-based initial weights (RBNN) have been compared in Figure 4.3. The error between analytical and RBNN results is plotted in Figure 4.4.

Example 4.2

Let us consider a first-order ODE

$$\frac{dy}{dx}+\left(x+\frac{1+3x^2}{1+x+x^3}\right)y = x^3+2x+x^2\left(\frac{1+3x^2}{1+x+x^3}\right) \quad x\in[0,1]$$

with initial condition $y(0)=1$.

The trial solution is the same as in Example 4.1.

We have trained the network for 20 equidistant points in [0, 1] and compared analytical and neural results with arbitrary and regression-based weights, with four, five, and six nodes fixed in the hidden layer. A comparison between analytical and neural results with arbitrary and regression-based weights is given in Table 4.2. Analytical results are included in the second column.

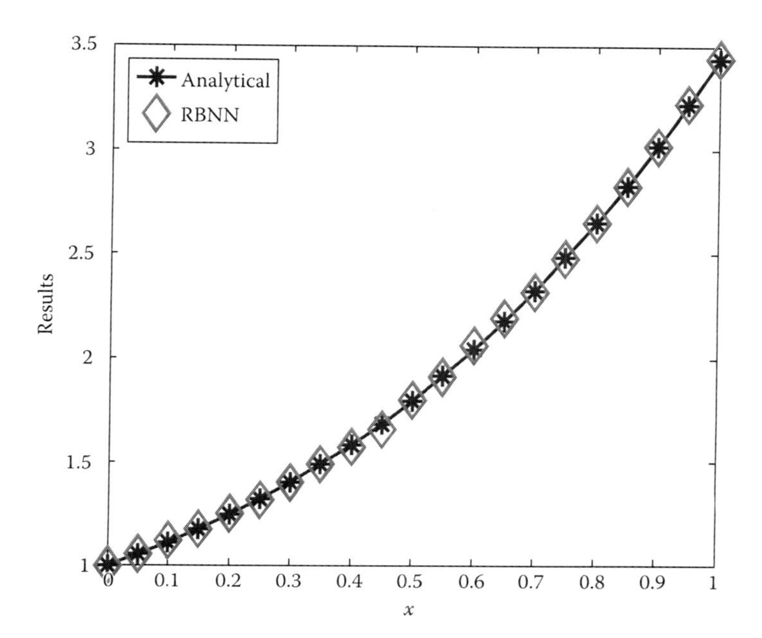

FIGURE 4.3
Plot of analytical and RBNN results (Example 4.1).

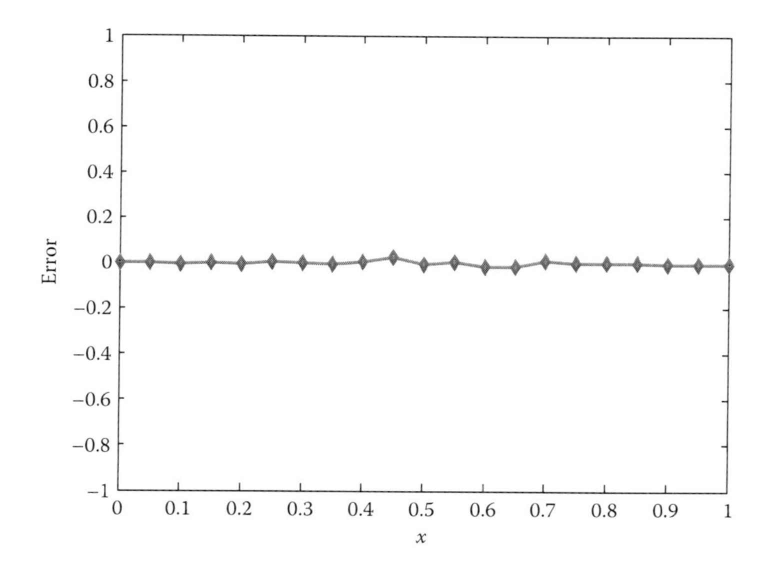

FIGURE 4.4
Error plot between analytical and RBNN results (Example 4.1).

TABLE 4.2
Analytical and Neural Results for All Combinations of Arbitrary and Regression-Based Weights (Example 4.2)

		Neural Results							
Input Data	Analytical	$w(A)$, $v(A)$ (Four Nodes)	$w(R)$, $v(R)$ RBNN (Four Nodes)	$w(A)$, $v(A)$ (Five Nodes)	$w(R)$, $v(R)$ RBNN (Five Nodes)	$w(A)$, $v(A)$ (Six Nodes)	Deviation (%)	$w(R)$, $v(R)$ RBNN (Six Nodes)	Deviation(%)
0	1.0000	1.0000	1.0000	1.0000	1.0000	1.0000	0.00	1.0000	0.00
0.05	0.9536	1.0015	0.9998	1.0002	0.9768	0.9886	3.67	0.9677	1.47
0.10	0.9137	0.9867	0.9593	0.9498	0.9203	0.9084	0.58	0.9159	0.24
0.15	0.8798	0.9248	0.8986	0.8906	0.8802	0.8906	1.22	0.8815	0.19
0.20	0.8514	0.9088	0.8869	0.8564	0.8666	0.8587	0.85	0.8531	0.19
0.25	0.8283	0.8749	0.8630	0.8509	0.8494	0.8309	0.31	0.8264	0.22
0.30	0.8104	0.8516	0.8481	0.8213	0.9289	0.8013	1.12	0.8114	0.12
0.35	0.7978	0.8264	0.8030	0.8186	0.8051	0.7999	0.26	0.7953	0.31
0.40	0.7905	0.8137	0.7910	0.8108	0.8083	0.7918	0.16	0.7894	0.13
0.45	0.7889	0.7951	0.7908	0.8028	0.7948	0.7828	0.77	0.7845	0.55
0.50	0.7931	0.8074	0.8063	0.8007	0.7960	0.8047	1.46	0.7957	0.32
0.55	0.8033	0.8177	0.8137	0.8276	0.8102	0.8076	0.53	0.8041	0.09
0.60	0.8200	0.8211	0.8190	0.8362	0.8246	0.8152	0.58	0.8204	0.04
0.65	0.8431	0.8617	0.8578	0.8519	0.8501	0.8319	1.32	0.8399	0.37
0.70	0.8731	0.8896	0.8755	0.8685	0.8794	0.8592	1.59	0.8711	0.22
0.75	0.9101	0.9281	0.9231	0.9229	0.9139	0.9129	0.31	0.9151	0.54
0.80	0.9541	0.9777	0.9613	0.9897	0.9603	0.9755	2.24	0.9555	0.14
0.85	1.0053	1.0819	0.9930	0.9956	1.0058	1.0056	0.03	0.9948	1.04
0.90	1.0637	1.0849	1.1020	1.0714	1.0663	1.0714	0.72	1.0662	0.23
0.95	1.1293	1.2011	1.1300	1.1588	1.1307	1.1281	0.11	1.1306	0.11
1.00	1.2022	1.2690	1.2195	1.2806	1.2139	1.2108	0.71	1.2058	0.29

Neural results for arbitrary weights $w(A)$ (from the input layer to the hidden layer) and $v(A)$ (from the hidden layer to the output layer) with four, five, and six nodes are presented in the third, fifth, and seventh column, respectively. Similarly, neural results with regression-based weights $w(R)$ (from the input layer to the hidden layer) and $v(R)$ (from the hidden layer to the output layer) with four, five, and six nodes are presented in the fourth, sixth, and ninth column, respectively.

Analytical and neural results with arbitrary and regression-based weights for six nodes in the hidden layer are compared in Figures 4.5 and 4.6, respectively. The error plot is shown in Figure 4.7. Absolute deviations in percent

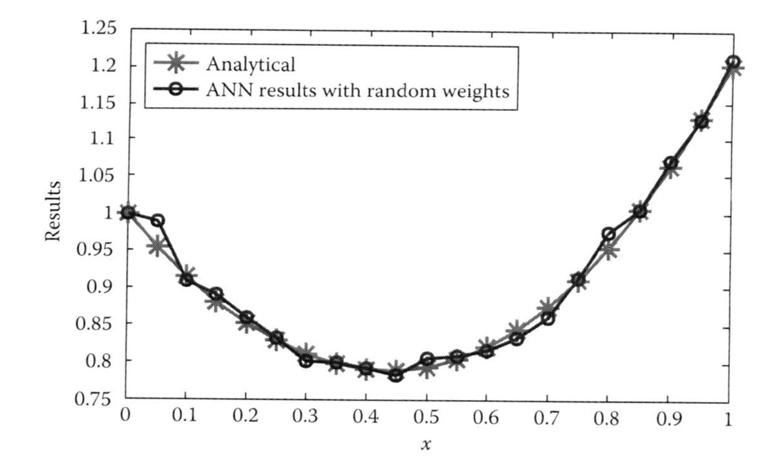

FIGURE 4.5
Plot of analytical and neural results with arbitrary weights (Example 4.2).

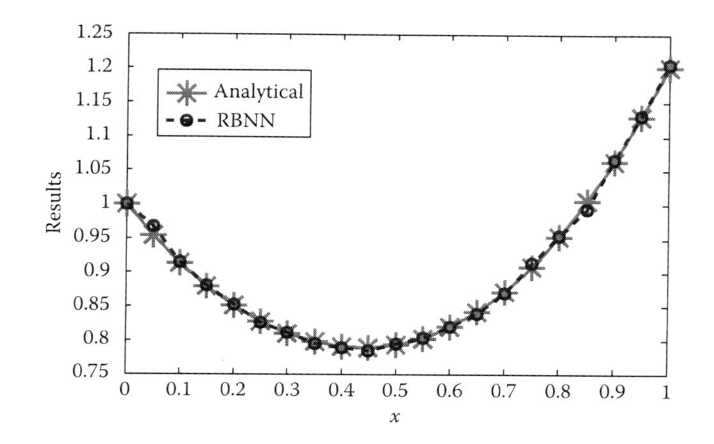

FIGURE 4.6
Plot of analytical and RBNN results for six nodes (Example 4.2).

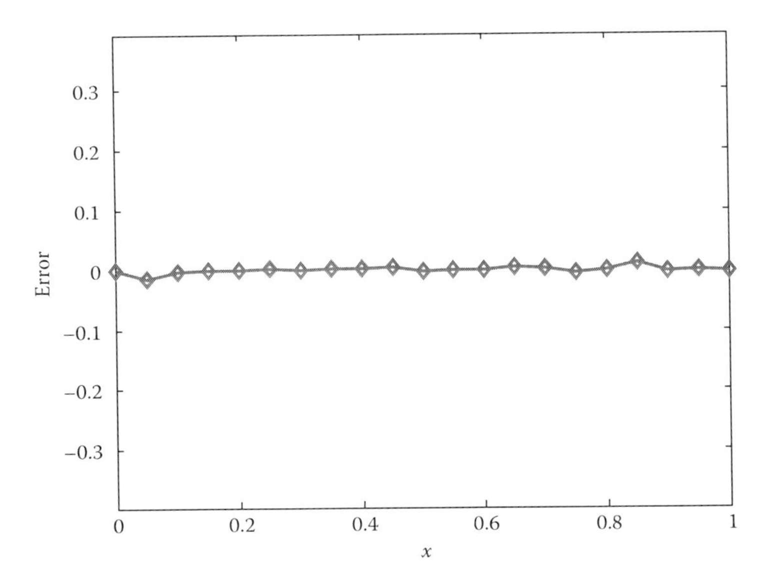

FIGURE 4.7
Error plot between analytical and RBNN results for six nodes (Example 4.2).

values have been calculated in Table 4.2, and the maximum deviation for neural results with arbitrary weights (six hidden nodes) is 3.67 (8th column), and for neural results with regression-based weights, it is 1.47 (10th column). From Figures 4.5 and 4.6, one can observe that neural results with regression-based weights exactly agree at all points with the analytical results, but for neural results with arbitrary weights, this is not the case. Thus, neural results with regression-based weights are more accurate.

It can be observed that by increasing the number of nodes in the hidden layer from four to six, the results are found to be better. However, when the number of nodes in the hidden layer is increased beyond six, the results do not improve further.

This problem has also been solved by well-known numerical methods, namely, Euler and Runge–Kutta for the sake of comparison. Table 4.3 shows the validation of the neural results (with six hidden nodes) by comparing with other numerical results (Euler and Runge–Kutta results).

Example 4.3

In this example, a first-order IVP has been considered

$$\frac{dy}{dx} = \alpha y$$

with initial condition $y(0) = 1$.

TABLE 4.3

Comparison of the Results (Example 4.2)

Input Data	Analytical	Euler	Runge–Kutta	$w(R)$, $v(R)$ RBNN (Six Nodes)
0	1.0000	1.0000	1.0000	1.0000
0.0500	0.9536	0.9500	0.9536	0.9677
0.1000	0.9137	0.9072	0.9138	0.9159
0.1500	0.8798	0.8707	0.8799	0.8815
0.2000	0.8514	0.8401	0.8515	0.8531
0.2500	0.8283	0.8150	0.8283	0.8264
0.3000	0.8104	0.7953	0.8105	0.8114
0.3500	0.7978	0.7810	0.7979	0.7953
0.4000	0.7905	0.7721	0.7907	0.7894
0.4500	0.7889	0.7689	0.7890	0.7845
0.5000	0.7931	0.7717	0.7932	0.7957
0.5500	0.8033	0.7805	0.8035	0.8041
0.6000	0.8200	0.7958	0.8201	0.8204
0.6500	0.8431	0.8178	0.8433	0.8399
0.7000	0.8731	0.8467	0.8733	0.8711
0.7500	0.9101	0.8826	0.9102	0.9151
0.8000	0.9541	0.9258	0.9542	0.9555
0.8500	1.0053	0.9763	1.0054	0.9948
0.9000	1.0637	1.0342	1.0638	1.0662
0.9500	1.1293	1.0995	1.1294	1.1306
1.000	1.2022	1.1721	1.2022	1.2058

This equation represents exponential growth, where $1/\alpha$ represents the time constant or characteristic time.

Considering $\alpha = 1$, we have the analytical solution as $y = e^x$.

The RBNN trial solution in this case is

$$y_t(x, p) = 1 + xN(x, p).$$

Now the network is trained for 10 equidistant points in the domain [0, 1] and 4, 5, and 6 hidden nodes are fixed according to regression-based algorithm. Next, 4, 5, and 6 hidden nodes are also taken for the traditional ANN model with random (arbitrary) initial weights. A comparison of analytical and neural results with arbitrary ($w(A)$, $v(A)$) and regression-based weights ($w(R)$, $v(R)$) have been given in Table 4.4. Analytical and traditional neural results obtained using random initial weights with six nodes are shown in Figure 4.8. Figure 4.9 depicts a comparison between analytical and neural results with regression-based initial weights for six hidden nodes.

TABLE 4.4

Analytical and Neural Results for All Combinations of Arbitrary and Regression-Based Weights (Example 4.3)

		Neural Results					
Input Data	Analytical	$w(A)$, $v(A)$ (Four Nodes)	$w(R)$, $v(R)$ RBNN (Four Nodes)	$w(A)$, $v(A)$ (Five Nodes)	$w(R)$, $v(R)$ RBNN (Five Nodes)	$w(A)$, $v(A)$ (Six Nodes)	$w(R)$, $v(R)$ RBNN (Six Nodes)
0	1.0000	1.0000	1.0000	1.0000	1.0000	1.0000	1.0000
0.1000	1.1052	1.1069	1.1061	1.1093	1.1060	1.1075	1.1051
0.2000	1.2214	1.2337	1.2300	1.2250	1.2235	1.2219	1.2217
0.3000	1.3499	1.3543	1.3512	1.3600	1.3502	1.3527	1.3498
0.4000	1.4918	1.4866	1.4921	1.4930	1.4928	1.4906	1.4915
0.5000	1.6487	1.6227	1.6310	1.6412	1.6456	1.6438	1.6493
0.6000	1.8221	1.8303	1.8257	1.8205	1.8245	1.8234	1.8220
0.7000	2.0138	2.0183	2.0155	2.0171	2.0153	2.0154	2.0140
0.8000	2.2255	2.2320	2.2302	2.2218	2.2288	2.2240	2.2266
0.9000	2.4596	2.4641	2.4625	2.4664	2.4621	2.4568	2.4597
1.0000	2.7183	2.7373	2.7293	2.7232	2.7177	2.7111	2.7186

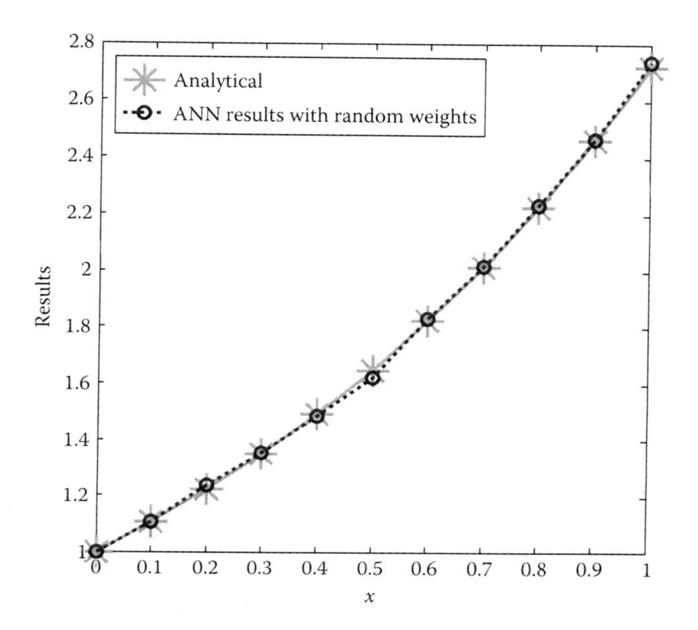

FIGURE 4.8
Plot of analytical and neural results with arbitrary weights for six nodes (Example 4.3).

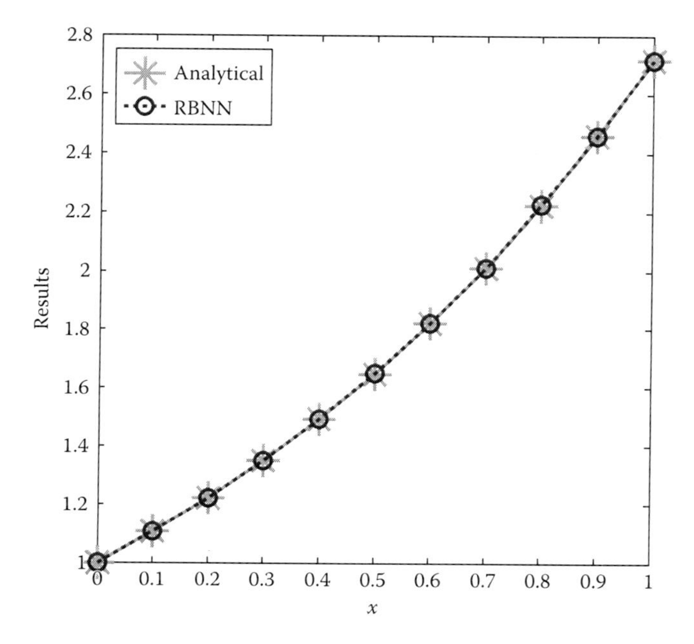

FIGURE 4.9
Plot of analytical and RBNN results for six nodes (Example 4.3).

4.6 Higher-Order Linear ODEs

In this section, one second-order IVP and one fourth-order BVP are taken in Examples 4.4 and 4.5, respectively.

Example 4.4

In this example, a second-order damped free vibration equation is taken as

$$\frac{d^2y}{dx^2}+4\frac{dy}{dx}+4y=0 \quad x\in[0,4]$$

with initial conditions $y(0)=1$, $y'(0)=1$.

As discussed in Section 3.2.2.2 (Equation 3.17), we can write the trial solution as

$$y_t(x,p)=1+x+x^2N(x,p).$$

Here, the network is trained for 40 equidistant points in [0, 4] and four, five, and six hidden nodes are fixed according to regression-based algorithm. In Table 4.5, we compare analytical results with neural results, taking arbitrary and regression-based weights for four, five, and six nodes in the hidden layer. Neural results for arbitrary weights $w(A)$ (from the input layer to the hidden layer) and $v(A)$ (from the hidden layer to the output layer) with four, five, and six nodes are shown in the third, fifth, and seventh column, respectively. Neural results with regression-based weights $w(R)$ (from the input layer to the hidden layer) and $v(R)$ (from the hidden layer to the output layer) with four, five, and six nodes are presented in the fourth, sixth, and eighth column, respectively.

Analytical and neural results obtained for random initial weights are depicted in Figure 4.10. Figure 4.11 shows a comparison between analytical and neural results for regression-based initial weights for six hidden nodes. Finally, the error plot between analytical and RBNN results is shown in Figure 4.12.

Example 4.5

Now, we solve a fourth-order ODE

$$\frac{d^4y}{dx^4}=120x \quad x\in[-1,1]$$

with boundary conditions $y(-1)=1$, $y(1)=3$, $y'(-1)=5$, $y'(1)=5$.

The ANN trial solution in this case is represented as (Section 3.2.3.2, Equation 3.28)

$$y_t(x,p)=-2x^4+2x^3+4x^2-x+(x+1)^2(x-1)^2N(x,p).$$

TABLE 4.5

Analytical and Neural Results for All Combinations of Arbitrary and Regression-Based Weights (Example 4.4)

Input Data	Analytical	Neural Results					
		$w(A), v(A)$ (Four Nodes)	$w(R), v(R)$ RBNN (Four Nodes)	$w(A), v(A)$ (Five Nodes)	$w(R), v(R)$ RBNN (Five Nodes)	$w(A), v(A)$ (Six Nodes)	$w(R), v(R)$ RBNN (Six Nodes)
0	1.0000	1.0000	1.0000	1.0000	1.0000	1.0000	1.0000
0.1	1.0643	1.0900	1.0802	1.0910	1.0878	1.0923	1.0687
0.2	1.0725	1.1000	1.0918	1.0858	1.0715	1.0922	1.0812
0.3	1.0427	1.0993	1.0691	1.0997	1.0518	1.0542	1.0420
0.4	0.9885	0.9953	0.9732	0.9780	0.9741	0.8879	0.9851
0.5	0.9197	0.9208	0.9072	0.9650	0.9114	0.9790	0.9122
0.6	0.8433	0.8506	0.8207	0.8591	0.8497	0.8340	0.8082
0.7	0.7645	0.7840	0.7790	0.7819	0.7782	0.7723	0.7626
0.8	0.6864	0.7286	0.6991	0.7262	0.6545	0.6940	0.6844
0.9	0.6116	0.6552	0.5987	0.6412	0.6215	0.6527	0.6119
1.0	0.5413	0.5599	0.5467	0.5604	0.5341	0.5547	0.5445
1.1	0.4765	0.4724	0.4847	0.4900	0.4755	0.4555	0.4634
1.2	0.4173	0.4081	0.4035	0.4298	0.4202	0.4282	0.4172
1.3	0.3639	0.3849	0.3467	0.3907	0.3761	0.3619	0.3622
1.4	0.3162	0.3501	0.3315	0.3318	0.3274	0.3252	0.3100
1.5	0.2738	0.2980	0.2413	0.2942	0.2663	0.2773	0.2759
1.6	0.2364	0.2636	0.2507	0.2620	0.2439	0.2375	0.2320
1.7	0.2036	0.2183	0.2140	0.2161	0.2107	0.2177	0.1921
1.8	0.1749	0.2018	0.2007	0.1993	0.1916	0.1622	0.1705
1.9	0.1499	0.1740	0.1695	0.1665	0.1625	0.1512	0.1501
2.0	0.1282	0.1209	0.1204	0.1371	0.1299	0.1368	0.1245

(Continued)

TABLE 4.5 (*Continued*)

Analytical and Neural Results for All Combinations of Arbitrary and Regression-Based Weights (Example 4.4)

		Neural Results					
Input Data	Analytical	$w(A), v(A)$ (Four Nodes)	$w(R), v(R)$ RBNN (Four Nodes)	$w(A), v(A)$ (Five Nodes)	$w(R), v(R)$ RBNN (Five Nodes)	$w(A), v(A)$ (Six Nodes)	$w(R), v(R)$ RBNN (Six Nodes)
2.1	0.1095	0.1236	0.1203	0.1368	0.1162	0.1029	0.1094
2.2	0.0933	0.0961	0.0942	0.0972	0.0949	0.0855	0.09207
2.3	0.0794	0.0818	0.0696	0.0860	0.0763	0.0721	0.0761
2.4	0.0675	0.0742	0.0715	0.0849	0.0706	0.0526	0.0640
2.5	0.0573	0.0584	0.0419	0.0609	0.0543	0.0582	0.0492
2.6	0.0485	0.0702	0.0335	0.0533	0.0458	0.0569	0.0477
2.7	0.0411	0.0674	0.0602	0.0581	0.0468	0.0462	0.0409
2.8	0.0348	0.0367	0.0337	0.0387	0.0328	0.0357	0.03460
2.9	0.0294	0.0380	0.0360	0.0346	0.0318	0.0316	0.0270
3.0	0.0248	0.0261	0.0207	0.0252	0.0250	0.0302	0.0247
3.1	0.0209	0.0429	0.0333	0.0324	0.0249	0.0241	0.0214
3.2	0.0176	0.0162	0.0179	0.0154	0.0169	0.0166	0.0174
3.3	0.0148	0.0159	0.0137	0.0158	0.0140	0.0153	0.0148
3.4	0.0125	0.0138	0.0135	0.0133	0.0130	0.0133	0.0129
3.5	0.0105	0.0179	0.0167	0.0121	0.0132	0.0100	0.0101
3.6	0.0088	0.0097	0.0096	0.0085	0.0923	0.0095	0.0090
3.7	0.0074	0.0094	0.0092	0.0091	0.0093	0.0064	0.0071
3.8	0.0062	0.0081	0.0078	0.0083	0.0070	0.0061	0.0060
3.9	0.0052	0.0063	0.0060	0.0068	0.0058	0.0058	0.0055
4.0	0.0044 .	0.0054	0.0052	0.0049	0.0049	0.0075	0.0046

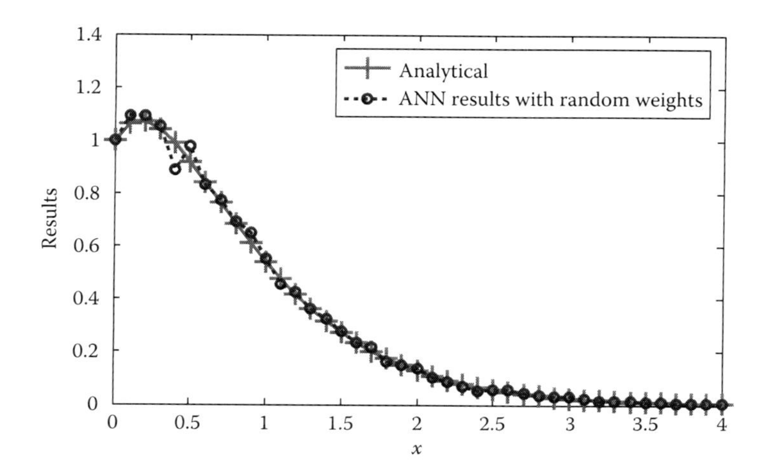

FIGURE 4.10
Plot of analytical and neural results with arbitrary weights (for six nodes) (Example 4.4).

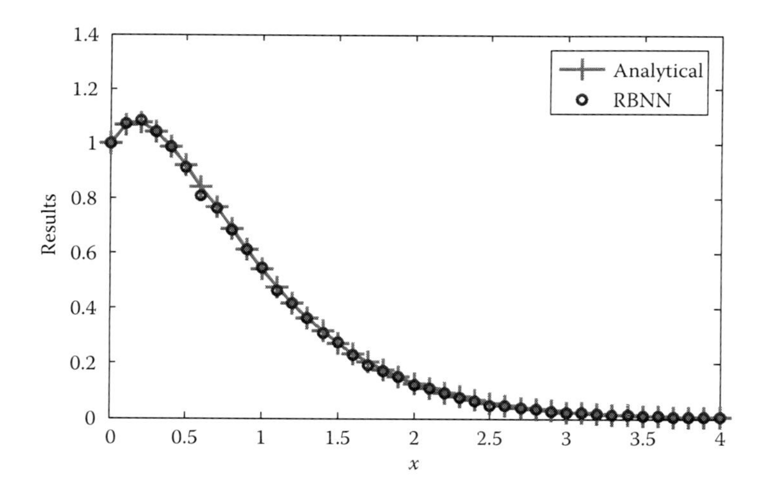

FIGURE 4.11
Plot of analytical and RBNN results for six nodes (Example 4.4).

The network has been trained for eight equidistant points in $[-1, 1]$ and four hidden nodes are fixed according to regression analysis. We have taken six nodes in the hidden layer for the traditional ANN model. Here, the tangent hyperbolic function is considered as the activation function. As in the previous case, analytical and obtained neural results with random initial weights are shown in Figure 4.13. A comparison between analytical and neural results for regression-based initial weights is depicted in Figure 4.14. Lastly, the error (between analytical and RBNN results) is plotted in Figure 4.15.

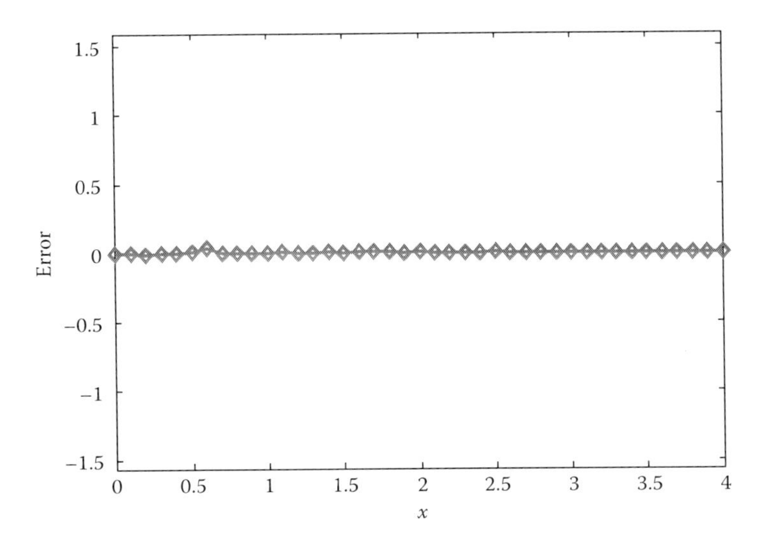

FIGURE 4.12
Error plot between analytical and RBNN solutions for six nodes (Example 4.4).

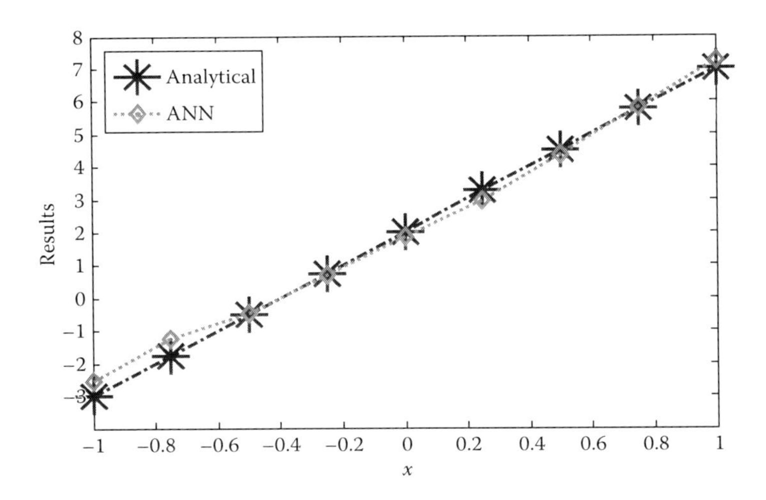

FIGURE 4.13
Plot of analytical and neural results with arbitrary weights (Example 4.5).

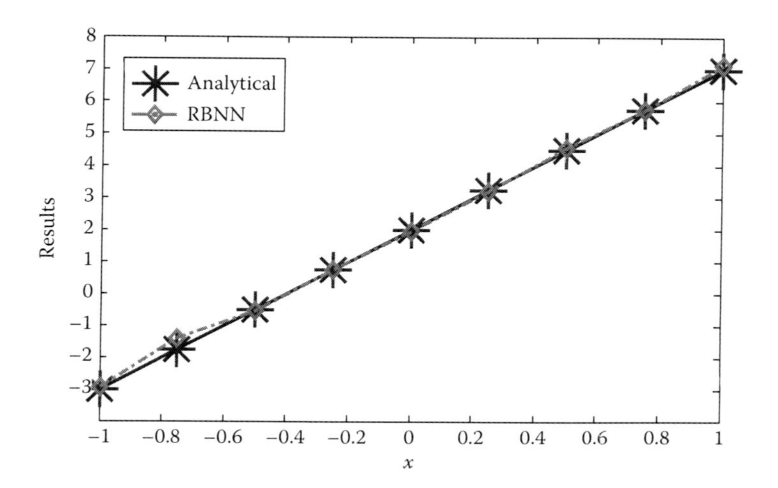

FIGURE 4.14
Plot of analytical and RBNN results (Example 4.5).

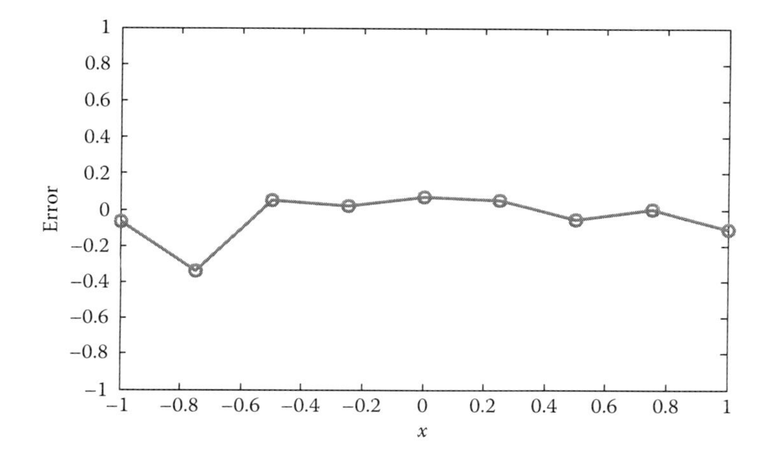

FIGURE 4.15
Error plot between analytical and RBNN results (Example 4.5).

References

1. S. Mall and S. Chakraverty. Comparison of artificial neural network architecture in solving ordinary differential equations. *Advances in Artificial Neural Systems*, 2013: 1–24, October 2013.
2. S. Mall and S. Chakraverty. Regression-based neural network training for the solution of ordinary differential equations. *International Journal of Mathematical Modelling and Numerical Optimization*, 4(2): 136–149, 2013.
3. S. Chakraverty and S. Mall. Regression based weight generation algorithm in neural network for solution of initial and boundary value problems. *Neural Computing and Applications*, 25(3): 585–594, September 2014.
4. S. Mall and S. Chakraverty. Regression based neural network model for the solution of initial value problem. *National Conference on Computational and Applied Mathematics in Science and Engineering* (*CAMSE-2012*), VNIT, Nagpur, India, December 2012.
5. S. Chakraverty, V.P. Singh, and R.K. Sharma. Regression based weight generation algorithm in neural network for estimation of frequencies of vibrating plates. *Journal of Computer Methods in Applied Mechanics and Engineering*, 195: 4194–4202, July 2006.
6. S. Chakraverty, V.P. Singh, R.K. Sharma, and G.K. Sharma. Modelling vibration frequencies of annular plates by regression based neural Network. *Applied Soft Computing*, 9(1): 439–447, January 2009.
7. E. Lagaris, A. Likas, and D.I. Fotiadis. Artificial neural networks for solving ordinary and partial differential equations. *IEEE Transactions on Neural Networks*, 9(5): 987–1000, September 1998.
8. D.R. Parisi, M.C. Mariani, and M.A. Laborde. Solving differential equations with unsupervised neural networks. *Chemical Engineering and Processing*, 42: 715–721, 2003.
9. S.T. Karris. *Numerical Analysis: Using MATLAB and Spreadsheets*. Orchard Publication, Fremont, CA, 2004.
10. R.B. Bhat and S. Chakraverty. *Numerical Analysis in Engineering*. Alpha Science Int., Ltd., Oxford, U.K., 2004.

5

Single-Layer Functional Link Artificial Neural Network

In this chapter, we introduce various types of functional link artificial neural network (FLANN) models to handle ordinary differential equations (ODEs) [1–5]. FLANN models are fast-learning single-layer artificial neural network (ANN) models. The single-layer FLANN method is introduced by Pao and Philips [6]. In FLANN, the hidden layer is replaced by a functional expansion block for enhancement of the input patterns using orthogonal polynomials such as Chebyshev, Legendre, Hermite, etc. In FLANN, the total number of network parameters is less than that of the multilayer perceptron (MLP) structure. So the technique is computationally more efficient than MLP.

In view of this discussion, some of the advantages of the single-layer FLANN-based model for solving differential equations may be mentioned as follows:

1. It is a single-layer neural network, so the total number of network parameters is less than that of traditional multilayer ANN.
2. It is capable of fast learning and is computationally efficient.
3. It is simple to implement and easy to compute.
4. Hidden layers are not required.
5. The back propagation algorithm is unsupervised.
6. No optimization technique is to be used.

The architecture of the FLANN models consists of two parts: the first one is the numerical transformation part, and the second part is the learning part. In the numerical transformation part, each input data is expanded to several terms using orthogonal polynomials, namely, Chebyshev, Legendre, Hermite, etc. So orthogonal polynomials can be viewed as new input vectors.

A learning algorithm is used for updating the network parameters and minimizing the error function. Here, the unsupervised error back-propagation algorithm is used to update the network parameters (weights) from the input layer to the output layer. The nonlinear tangent hyperbolic (*tanh* viz. $(e^x - e^{-x})/(e^x + e^{-x})$) function is considered as the activation function. The gradient descent algorithm is used for learning, and the weights are updated by taking a negative gradient at each iteration. The weights (w_j) are initialized randomly and then updated as follows [7]:

$$w_j^{k+1} = w_j^k + \Delta w_j^k = w_j^k + \left(-\eta \frac{\partial E(x,p)}{\partial w_j^k} \right) \tag{5.1}$$

where

- η is the learning parameter (which lies between 0 and 1)
- k is the iteration step used to update the weights as usual in ANN
- $E(x,p)$ is the error function

As regards the Chebyshev neural network (ChNN), it has been applied to various problems, namely, system identification [8–10], digital communication [11], channel equalization [12], function approximation [13], etc. Recently, Mall and Chakraverty [1,2] developed the ChNN model for solving second-order singular initial value problems, namely, Lane–Emden and Emden–Fowler type equations.

Similarly, the single-layer Legendre neural network (LeNN) has been introduced by Yang and Tseng [14] for function approximation. Further, the LeNN model has been used for channel equalization problems [15,16], system identification [17], prediction of machinery noise [18], and differential equations [3]. In [4], Mall and Chakraverty compared the results of singular nonlinear ODEs using LeNN and ChNN. Very recently, they [5] proposed the Hermite neural network (HeNN) model also to handle Van der Pol–Duffing oscillator equations.

5.1 Single-Layer FLANN Models

In this section, we describe the structure of single-layer FLANN models, and then the formulations and gradient computations are explained.

5.1.1 ChNN Model

5.1.1.1 Structure of the ChNN Model

Figure 5.1 shows the structure of the ChNN model, which consists of a single input node, a functional expansion block based on Chebyshev polynomials, and a single output node. Here, the dimension of the input data is expanded using a set of Chebyshev polynomials. Let us denote the input data as $x = (x_1, x_2, \ldots, x_h)^T$, where the single input node x has h number of data. Chebyshev polynomials are a set of orthogonal polynomials obtained by the

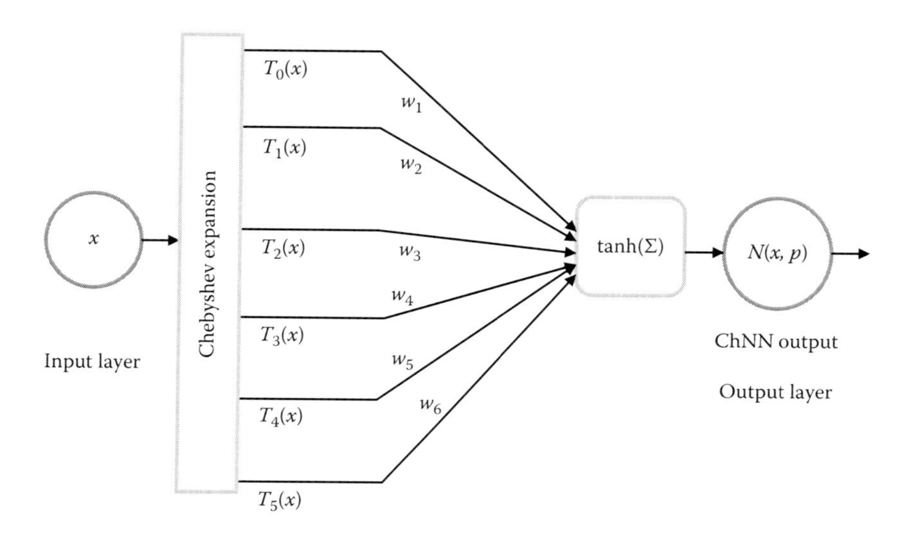

FIGURE 5.1
Structure of single-layer ChNN.

solution of Chebyshev differential equations. The first two Chebyshev polynomials are known as

$$\left.\begin{aligned} T_0(x) &= 1 \\ T_1(x) &= x \end{aligned}\right\} \tag{5.2}$$

Higher-order Chebyshev polynomials may be generated by the well-known recursive formula [19,20]

$$T_{r+1}(x) = 2xT_r(x) - T_{r-1}(x) \tag{5.3}$$

where $T_r(x)$ denotes the rth-order Chebyshev polynomial. Here, an n-dimensional input data is expanded to m-dimensional enhanced Chebyshev polynomials. The advantage of ChNN is that it gets the result by using a single-layer network, although this is done by increasing the dimension of the input data through the Chebyshev polynomial.

5.1.1.2 Formulation of the ChNN Model

The general formulation of ODEs using ANN is discussed in Section 3.2.1.

The ChNN trial solution $y_t(x, p)$ for ODEs with parameters (weights) p may be written in the form

$$y_t(x, p) = A(x) + F(x, N(x, p)) \tag{5.4}$$

The first term $A(x)$ does not contain adjustable parameters and satisfies only initial/boundary conditions, whereas the second term $F(x,N(x,p))$ contains the single output $N(x,p)$ of ChNN with input x and adjustable parameters p. The tangent hyperbolic (tanh) function, namely, $(e^x - e^{-x})/(e^x + e^{-x})$, is considered here as the activation function.

The network output with input x and parameters (weights) p may be computed as

$$N(x,p) = \tanh(z) = \frac{e^z - e^{-z}}{e^z + e^{-z}} \tag{5.5}$$

where z is a weighted sum of the expanded input data. It is written as

$$z = \sum_{j=1}^{m} w_j T_{j-1}(x) \tag{5.6}$$

where

$x = (x_1, x_2, \ldots, x_h)^T$ denotes the input data

$T_{j-1}(x)$ and w_j, with $j = \{1,2,3, \ldots, m\}$, are the expanded input data and the weight vectors, respectively, of the ChNN model

For minimizing the error function $E(x,p)$ discussed in Section 3.2.2 (Equation 3.7), that is, to update the network parameters (weights), we differentiate $E(x,p)$ with respect to the parameters. Then, the gradient of the network output with respect to its input is computed as discussed next.

5.1.1.3 Gradient Computation of the ChNN Model

Error computation not only involves the output but also the derivative of the network output with respect to its input. Thus, it requires finding out the gradient of the network derivatives with respect to their input.

As such, the derivative of $N(x,p)$ with respect to input x is written as

$$\frac{dN}{dx} = \sum_{j=1}^{m} \left[\left(\frac{\left(e^{(w_j T_{j-1}(x))} + e^{-(w_j T_{j-1}(x))}\right)^2 - \left(e^{(w_j T_{j-1}(x))} - e^{-(w_j T_{j-1}(x))}\right)^2}{\left(e^{(w_j T_{j-1}(x))} + e^{-(w_j T_{j-1}(x))}\right)^2} \right) \right] \left(w_j T'_{j-1}(x)\right) \tag{5.7}$$

Simplifying this, we have

$$\frac{dN}{dx} = \sum_{j=1}^{m} \left[1 - \left(\frac{e^{(w_j T_{j-1}(x))} - e^{-(w_j T_{j-1}(x))}}{e^{(w_j T_{j-1}(x))} + e^{-(w_j T_{j-1}(x))}} \right)^2 \right] \left(w_j T'_{j-1}(x)\right) \tag{5.8}$$

It may be noted that this differentiation is done for all x, where x has h number of data.

Similarly, we can compute the second derivative of $N(x, p)$ as

$$\frac{d^2N}{dx^2}=\sum_{j=1}^{m}\left[\begin{array}{l}\left\{-2\left(\frac{e^{(w_jT_{j-1}(x))}-e^{-(w_jT_{j-1}(x))}}{e^{(w_jT_{j-1}(x))}+e^{-(w_jT_{j-1}(x))}}\right)\right.\\ \left.\times\left(\frac{\left(e^{(w_jT_{j-1}(x))}+e^{-(w_jT_{j-1}(x))}\right)^2-\left(e^{(w_jT_{j-1}(x))}-e^{-(w_jT_{j-1}(x))}\right)^2}{\left(e^{(w_jT_{j-1}(x))}+e^{-(w_jT_{j-1}(x))}\right)^2}\right)\right\}(w_jT'_{j-1}(x))^2\\ +(w_jT''_{j-1}(x))\left\{1-\left(\frac{e^{(w_jT_{j-1}(x))}-e^{-(w_jT_{j-1}(x))}}{e^{(w_jT_{j-1}(x))}+e^{-(w_jT_{j-1}(x))}}\right)^2\right\}\end{array}\right] \tag{5.9}$$

After simplifying this, we get

$$\frac{d^2N}{dx^2}=\sum_{j=1}^{m}\left[\begin{array}{l}\left(\left\{2\left(\frac{e^{(w_jT_{j-1}(x))}-e^{-(w_jT_{j-1}(x))}}{e^{(w_jT_{j-1}(x))}+e^{-(w_jT_{j-1}(x))}}\right)^3-2\left(\frac{e^{(w_jT_{j-1}(x))}-e^{-(w_jT_{j-1}(x))}}{e^{(w_jT_{j-1}(x))}+e^{-(w_jT_{j-1}(x))}}\right)\right\}(w_jT'_{j-1}(x))^2\right)\\ +\left(\left\{1-\left(\frac{e^{(w_jT_{j-1}(x))}-e^{-(w_jT_{j-1}(x))}}{e^{(w_jT_{j-1}(x))}+e^{-(w_jT_{j-1}(x))}}\right)^2\right\}(w_jT''_{j-1}(x))\right)\end{array}\right] \tag{5.10}$$

where w_j denotes parameters of network and $T'_{j-1}(x), T''_{j-1}(x)$ denote first and second derivatives of Chebyshev polynomials.

Let $N_\delta = dN/dx$ denote the derivative of the network output with respect to input x. The derivative of $N(x,p)$ and N_δ with respect to other parameters (weights) may be formulated as

$$\frac{\partial N}{\partial w_j}=\sum_{j=1}^{m}\left[1-\left(\frac{e^{(w_jT_{j-1}(x))}-e^{-(w_jT_{j-1}(x))}}{e^{(w_jT_{j-1}(x))}+e^{-(w_jT_{j-1}(x))}}\right)^2\right](T_{j-1}(x)) \tag{5.11}$$

$$\frac{\partial N_\delta}{\partial w_j}=\left[(\tanh(z))''(T_{j-1}(x))(w_jT'_{j-1}(x))\right]+\left[(\tanh(z))'(T'_{j-1}(x))\right] \tag{5.12}$$

After getting all the derivatives, we can find out the gradient of error. Using the unsupervised error back-propagation learning algorithm (in Section 3.2.2,

Equations 3.8 and 3.10), we may minimize the error function as per the desired accuracy.

5.1.2 LeNN Model

5.1.2.1 Structure of the LeNN Model

Figure 5.2 depicts the structure of the single-layer LeNN model, which consists of a single input node, one output layer, and a functional expansion block based on Legendre polynomials (first five). The hidden layer is eliminated by transforming the input pattern to a higher-dimensional space using Legendre polynomials. The Legendre polynomials are denoted by $L_n(u)$; here, n is the order and $-1 < u < 1$ is the argument of the polynomial. Legendre polynomials constitute a set of orthogonal polynomials obtained as a solution of Legendre differential equations.

The first few Legendre polynomials are [20] as follows:

$$\left.\begin{aligned} L_0(u) &= 1 \\ L_1(u) &= u \\ L_2(u) &= \frac{1}{2}\left(3u^2 - 1\right) \end{aligned}\right\} \tag{5.13}$$

Higher-order Legendre polynomials may be generated by the following recursive formula:

$$L_{n+1}(u) = \frac{1}{n+1}\left[(2n+1)uL_n(u) - nL_{n-1}(u)\right] \tag{5.14}$$

As input data, we consider a vector $x = (x_1, x_2, \ldots, x_h)$ of dimension h.

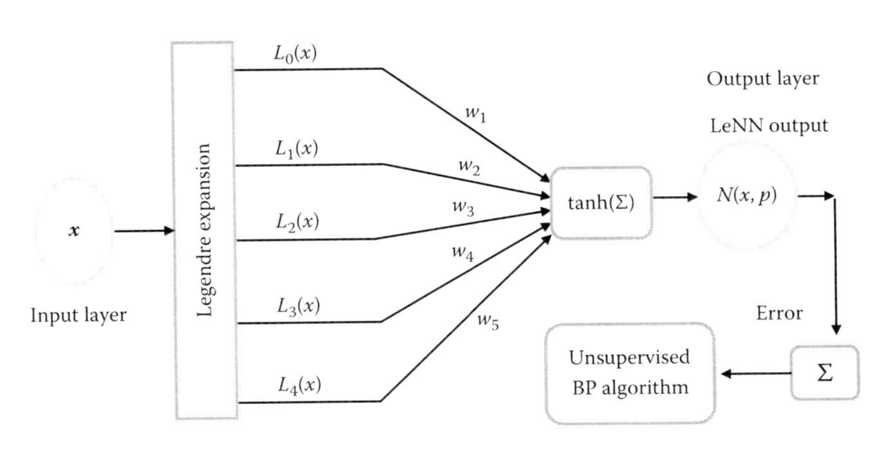

FIGURE 5.2
Structure of LeNN.

The enhanced pattern is obtained by using the Legendre polynomials

$$[1, L_1(x_1), L_2(x_1), L_3(x_1), \ldots, L_n(x_1); 1, L_1(x_2), L_2(x_2), L_3(x_2), \ldots, L_n(x_2); 1, L_1(x_h), L_2(x_h), L_3(x_h), \ldots, L_n(x_h)]$$

Here, h-dimensional data is expanded to n-dimensional enhanced Legendre polynomials.

5.1.2.2 Formulation of the LeNN Model

The LeNN trial solution for ODEs may be expressed as

$$y_t(x,p) = A(x) + F(x, N(x,p)) \tag{5.15}$$

where the first term $A(x)$ satisfies initial/boundary conditions. The second term, namely, $F(x,N(x,p))$, contains the single output $N(x,p)$ of the LeNN model with one input node x (having h number of data) and adjustable parameters p.

Here,

$$N(x,p) = \tanh(z) = \frac{e^z - e^{-z}}{e^z + e^{-z}} \tag{5.16}$$

and

$$z = \sum_{j=1}^{m} w_j L_{j-1}(x) \quad j = 1, 2, \ldots, m \tag{5.17}$$

where

x is the input data

$L_{j-1}(x)$ and w_j, for $j = \{1, 2, 3, \ldots, m\}$, denote the expanded input data and the weight vectors, respectively, of the LeNN model

The nonlinear tangent hyperbolic tanh (·) function is considered as the activation function.

5.1.2.3 Gradient Computation of the LeNN Model

For minimizing the error function $E(x,p)$, we differentiate $E(x,p)$ with respect to the network parameters. Thus, the gradient of the network output with respect to its input is computed as discussed next.

The derivatives of $N(x,p)$ with respect to input x is expressed as

$$\frac{dN}{dx}=\sum_{j=1}^{m}\left[\left(\frac{\left(e^{(w_jL_{j-1}(x))}+e^{-(w_jL_{j-1}(x))}\right)^2-\left(e^{(w_jL_{j-1}(x))}-e^{-(w_jL_{j-1}(x))}\right)^2}{\left(e^{(w_jL_{j-1}(x))}+e^{-(w_jL_{j-1}(x))}\right)^2}\right)\right]\left(w_jL'_{j-1}(x)\right) \tag{5.18}$$

Simplifying Equation 5.18, we have

$$\frac{dN}{dx}=\sum_{j=1}^{m}\left[1-\left(\frac{e^{(w_jL_{j-1}(x))}-e^{-(w_jL_{j-1}(x))}}{e^{(w_jL_{j-1}(x))}+e^{-(w_jL_{j-1}(x))}}\right)^2\right]\left(w_jL'_{j-1}(x)\right) \tag{5.19}$$

Similarly, we can compute the second derivative of $N(x, p)$ as

$$\frac{d^2N}{dx^2}=\sum_{j=1}^{m}\left[\begin{array}{l}\frac{d}{dx}\left\{1-\left(\frac{e^{(w_jL_{j-1}(x))}-e^{-(w_jL_{j-1}(x))}}{e^{(w_jL_{j-1}(x))}+e^{-(w_jL_{j-1}(x))}}\right)^2\right\}\left(w_jL'_{j-1}(x)\right)\\ +\frac{d}{dx}\left(w_jL'_{j-1}(x)\right)\left\{1-\left(\frac{e^{(w_jL_{j-1}(x))}-e^{-(w_jL_{j-1}(x))}}{e^{(w_jL_{j-1}(x))}+e^{-(w_jL_{j-1}(x))}}\right)^2\right\}\end{array}\right] \tag{5.20}$$

After simplifying this, we get

$$\frac{d^2N}{dx^2}=\sum_{j=1}^{m}\left[\begin{array}{l}\left(\left\{2\left(\frac{e^{(w_jL_{j-1}(x))}-e^{-(w_jL_{j-1}(x))}}{e^{(w_jL_{j-1}(x))}+e^{-(w_jL_{j-1}(x))}}\right)^3-2\left(\frac{e^{(w_jL_{j-1}(x))}-e^{-(w_jL_{j-1}(x))}}{e^{(w_jL_{j-1}(x))}+e^{-(w_jL_{j-1}(x))}}\right)\right\}\left(w_jL'_{j-1}(x)\right)^2\right)\\ +\left(\left\{1-\left(\frac{e^{(w_jL_{j-1}(x))}-e^{-(w_jL_{j-1}(x))}}{e^{(w_jL_{j-1}(x))}+e^{-(w_jL_{j-1}(x))}}\right)^2\right\}\left(w_jL''_{j-1}(x)\right)\right)\end{array}\right] \tag{5.21}$$

where w_j, $L'_{j-1}(x)$, and $L''_{j-1}(x)$ denote weights of network, and first and second derivatives of Legendre polynomials, respectively.

5.1.3 HeNN Model

5.1.3.1 Architecture of the HeNN Model

Figure 5.3 depicts the structure of the HeNN model, which consists of a single input node, a single output node, and a Hermite orthogonal polynomial (first seven)-based functional expansion block. The HeNN model is a single-layer neural model where each input data is expanded to several

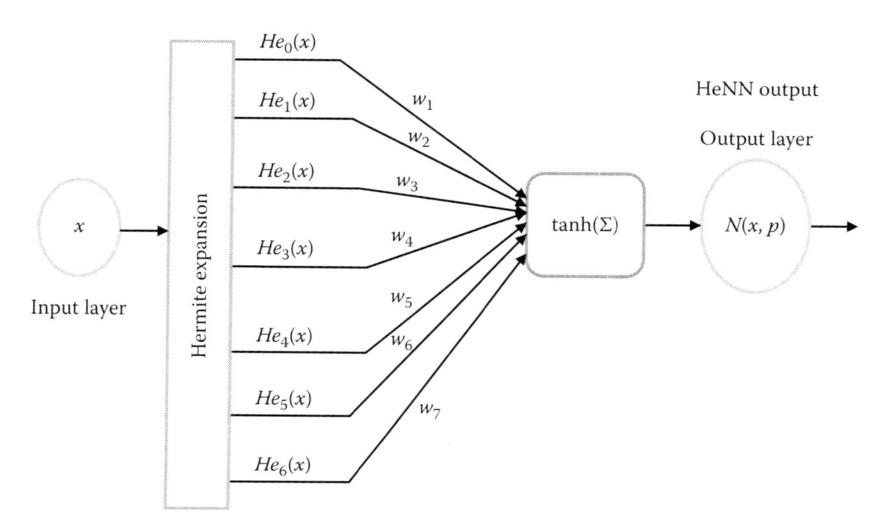

FIGURE 5.3
Architecture of single-layer HeNN.

terms using Hermite polynomials. The first three Hermite polynomials may be written as [21]

$$\left.\begin{aligned} He_0(x) &= 1 \\ He_1(x) &= x \\ He_2(x) &= x^2 - 1 \end{aligned}\right\} \tag{5.22}$$

Higher-order Hermite polynomials may then be generated by the recursive formula

$$He_{n+1}(x) = xHe_n(x) - He'_n(x) \tag{5.23}$$

We consider input data $x = (x_1, x_2, \ldots, x_h)$; that is, the single node x is assumed to have h number of data.

5.1.3.2 Formulation of the HeNN Model

The HeNN trial solution $y_{He}(x,p)$ for the ODEs with parameters (weights) p may be expressed as

$$y_{He}(x,p) = A(x) + G(x, N(x,p)) \tag{5.24}$$

The first term $A(x)$ does not contain adjustable parameters and satisfies only initial/boundary conditions, whereas the second term $G(x, N(x,p))$ contains the single output $N(x,p)$ of the HeNN model with input x and adjustable parameters p.

As such, the network output with input x and parameters (weights) p may be computed as

$$N(x,p) = \tanh(z) = \frac{e^{z} - e^{-z}}{e^{z} + e^{-z}} \tag{5.25}$$

where z is a linear combination of the expanded input data and is written as

$$z = \sum_{j=1}^{m} w_j He_{j-1}(x) \tag{5.26}$$

Here,

x is the input data

$He_{j-1}(x)$ and w_j, with $j = \{1, 2, 3, \ldots, m\}$, denote the expanded input data and the weight vectors, respectively, of the HeNN model.

5.1.4 Simple Orthogonal Polynomial–Based Neural Network (SOPNN) Model

5.1.4.1 Structure of the SOPNN Model

A single-layer simple orthogonal polynomial–based neural network model (SOPNN) has been considered here. Figure 5.4 gives the structure

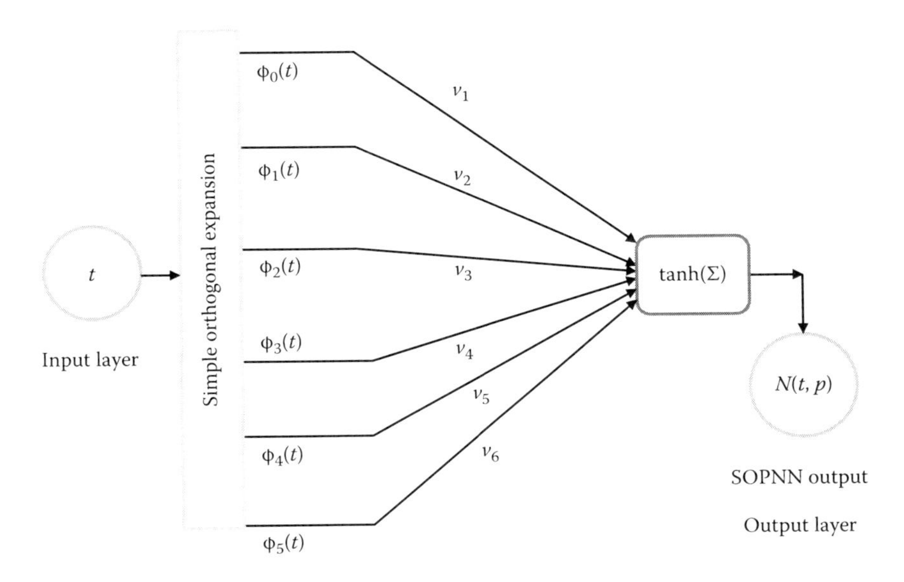

FIGURE 5.4
Structure of single-layer SOPNN.

of the SOPNN model, which consists of a single input node, a single output node, and a functional expansion block based on Gram–Schmidt orthogonal polynomials. In the numerical transformation part, each input data of the SOPNN model is expanded to several terms using Gram–Schmidt orthogonal polynomials. We have considered only one input node. We consider the input data as $t = (t_1, t_2, \ldots, t_h)^T$; that is, the single node t has h number of data. Here, an h-dimensional input data is expanded to m-dimensional enhanced orthogonal polynomials. For the linearly independent sequence $\{1, u, u^2, u^3, \ldots\}$, the first six orthogonal polynomials obtained by the Gram–Schmidt process are well known and may be written as [21]

$$\left.\begin{aligned} &\phi_0(u) = 1 \\ &\phi_1(u) = u - \frac{1}{2} \\ &\phi_2(u) = u^2 - u + \frac{1}{6} \\ &\phi_3(u) = u^3 - \frac{93}{2}u^2 + \frac{3}{5}u - \frac{1}{20} \\ &\phi_4(u) = u^4 - 2u^3 + \frac{9}{7}u^2 - \frac{2}{7}u + \frac{1}{70} \\ &\phi_5(u) = u^5 - \frac{5}{2}u^4 + \frac{20}{9}u^3 - \frac{5}{6}u^2 + \frac{5}{42}u - \frac{1}{252} \\ &\vdots \end{aligned}\right\} \tag{5.27}$$

5.1.4.2 Formulation of the SOPNN Model

The SOPNN trial solution $x_\phi(t,p)$ for ODEs with input t and parameters p may be expressed as

$$x_\phi(t,p) = A(t) + F(t, N(t,p)) \tag{5.28}$$

The first term $A(t)$ satisfies only initial/boundary conditions, whereas the second term $F(t, N(t,p))$ contains the single output $N(t,p)$ of SOPNN with input t and adjustable parameters (weights) p. The tangent hyperbolic function is considered here as the activation function.

As mentioned in Section 5.1.4.1, a single-layer SOPNN is considered with one input node, and the single output node $N(t,p)$ is formulated as

$$N(t,p) = \tanh(z) = \frac{e^z - e^{-z}}{e^z + e^{-z}} \tag{5.29}$$

where z a linear combination of the expanded input data. It is written as

$$z = \sum_{j=1}^{m} v_j \phi_{j-1}(t) \tag{5.30}$$

where
t is the input data
$\phi_{j-1}(t)$ and v_j, with $j = \{1, 2, 3, \ldots, m\}$, denote the expanded input data and weight vectors, respectively, of the SOPNN model

5.1.4.3 Gradient Computation of the SOPNN Model

Error computation not only involves the output but also the derivative of the network output with respect to its input. Thus, it requires finding out the gradient of the network derivatives with respect to their input.

As such, the derivative of $N(t, p)$ with respect to input t is written as

$$\frac{dN}{dt} = \sum_{j=1}^{m} \left[\left(\frac{\left(e^{(v_j\phi_{j-1}(t))} + e^{-(v_j\phi_{j-1}(t))}\right)^2 - \left(e^{(v_j\phi_{j-1}(t))} - e^{-(v_j\phi_{j-1}(t))}\right)^2}{\left(e^{(v_j\phi_{j-1}(t))} + e^{-(v_j\phi_{j-1}(t))}\right)^2} \right) \right] \left(v_j \phi'_{j-1}(t)\right) \tag{5.31}$$

Similarly, we can find other derivatives, as in Section 3.2.5, Equations 3.43 and 3.48. Using the unsupervised error back-propagation learning algorithm discussed in Section 3.2.2 (Equations 3.8 and 3.10), we may minimize the error function as per the desired accuracy.

5.2 First-Order Linear ODEs

Here, the first-order ODE has been solved using the single-layer ChNN model.

Example 5.1

Let us consider the following first-order ODE:

$$\frac{dy}{dx} = 4x^3 - 3x^2 + 2 \quad x \in [a, b]$$

subject to $y(0) = 0$.

TABLE 5.1

Analytical and ChNN Results (Example 5.1)

Input Data	Analytical	ChNN
0	0	0
0.1	0.1991	0.1989
0.2	0.3936	0.3940
0.3	0.5811	0.5812
0.4	0.7616	0.7614
0.5	0.9375	0.9380
0.6	1.1136	1.1126
0.7	1.2971	1.2968
0.8	1.4976	1.4978
0.9	1.7271	1.7240
1	2.0000	2.0010

As discussed in Section 3.2.2.1 (Equation 3.13), we can write the trial solution as

$$y_t(x,p) = xN(x,p)$$

The network is trained for 10 equidistant points in [0, 1] and the first 6 Chebyshev polynomials. A comparison of analytical and ChNN results has been shown in Table 5.1.

In view of the table, one may see that the analytical results compared very well with ChNN results.

5.3 Higher-Order ODEs

Here, we compare multilayer ANN, ChNN, and LeNN results for a second-order nonlinear ODE.

Example 5.2

Let us consider the second-order homogeneous Lane–Emden equation

$$\frac{d^2y}{dx^2} + \frac{2}{x}\frac{dy}{dx} - 2(2x^2+3)y = 0$$

with initial conditions $y(0)=1$, $y'(0)=0$.

As mentioned in Section 3.2.2.2 (Equation 3.17), we have the trial solution as

$$y_t(x,p) = 1 + x^2N(x,p)$$

TABLE 5.2

Comparison among Analytical, MLP, ChNN, and LeNN results (Example 5.2)

Input Data	Analytical [22]	MLP	ChNN	LeNN
0	1.0000	1.0000	1.0002	1.0002
0.1	0.9900	0.9914	0.9901	0.9907
0.2	0.9608	0.9542	0.9606	0.9602
0.3	0.9139	0.9196	0.9132	0.9140
0.4	0.8521	0.8645	0.8523	0.8503
0.5	0.7788	0.7710	0.7783	0.7754
0.6	0.6977	0.6955	0.6974	0.6775
0.7	0.6126	0.6064	0.6116	0.6125
0.8	0.5273	0.5222	0.5250	0.5304
0.9	0.4449	0.4471	0.4439	0.4490
1.0	0.3679	0.3704	0.3649	0.3696

Ten equidistant points in the domain [0, 1] and six hidden nodes for traditional MLP are considered. We have taken first six polynomials (Chebyshev, Legendre) for a single-layer neural network. Table 5.2 shows a comparison among numerical results obtained by analytical, traditional MLP, ChNN, and LeNN. Also, this comparison among analytical, traditional MLP, ChNN, and LeNN results is plotted in Figure 5.5.

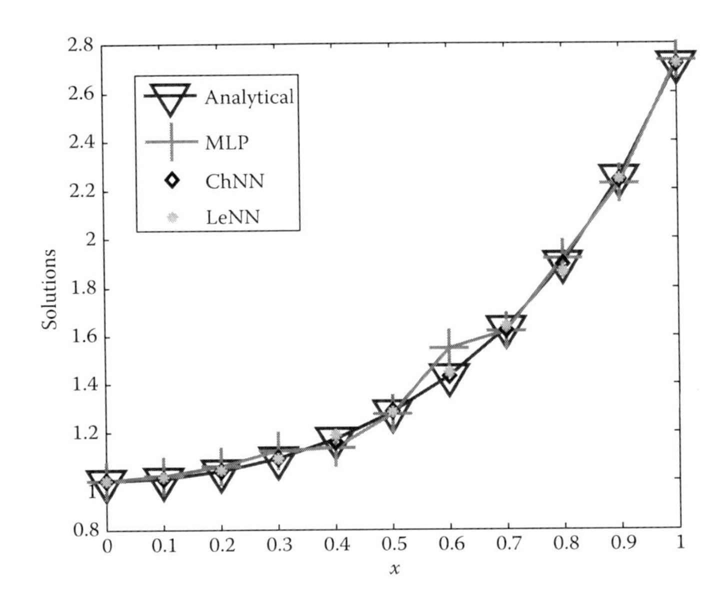

FIGURE 5.5
Comparison among Analytical, MLP, ChNN, and LeNN Results (Example 5.2).

The CPU time for computation of Example 5.2 for traditional ANN (MLP), ChNN, and LeNN is 10,168.41, 8,552.15, and 8,869.19 s, respectively. It can be observed that ChNN and LeNN require less time than MLP. Moreover, between ChNN and LeNN, ChNN requires less CPU time for computation of the present problem.

5.4 System of ODEs

Here, we present the solution of a system of ODEs using the single-layer LeNN method.

Example 5.3

In this example, a system of coupled first-order ODEs is taken

$$\left.\begin{aligned}\frac{dy_1}{dx} &= \frac{\cos(x)-\sin(x)}{y_2}\\ \frac{dy_2}{dx} &= y_1 y_2 + e^x - \sin(x)\end{aligned}\right\} \quad x \in [0,2]$$

subject to $y_1(0)=0$ and $y_2(0)=1$.

Analytical solutions for this may be obtained as

$$\left.\begin{aligned}y_1(x) &= \frac{\sin(x)}{e^x}\\ y_2(x) &= e^x\end{aligned}\right\}$$

Corresponding LeNN trial solutions are as follows:

$$\left.\begin{aligned}y_{t_1}(x) &= xN_1(x,p_1)\\ y_{t_2}(x) &= 1 + xN_2(x,p_2)\end{aligned}\right\}$$

Again, the network is trained here for 20 equidistant points in the given domain [0, 2]. As in previous cases, a comparison among analytical, LeNN, and MLP results is shown in Table 5.3. The comparison between analytical $(y_1(x), y_2(x))$ and LeNN $(y_{t_1}(x), y_{t_2}(x))$ results is plotted in Figure 5.6 and found to be in excellent agreement. A plot of the error function is presented in Figure 5.7. Last, LeNN results for some testing points are given in Table 5.4.

TABLE 5.3

Comparison among Analytical, LeNN, and MLP Results (Example 5.3)

Input Data	Analytical $y_1(x)$	LeNN $y_{t_1}(x)$	MLP $y_{t_1}(x)$	Analytical $y_2(x)$	LeNN $y_{t_2}(x)$	MLP $y_{t_2}(x)$
0	0	0	0	1.0000	1.0000	1.0000
0.1000	0.0903	0.0907	0.0899	1.1052	1.1063	1.1045
0.2000	0.1627	0.1624	0.1667	1.2214	1.2219	1.2209
0.3000	0.2189	0.2199	0.2163	1.3499	1.3505	1.3482
0.4000	0.2610	0.2609	0.2625	1.4918	1.5002	1.4999
0.5000	0.2908	0.2893	0.2900	1.6487	1.6477	1.6454
0.6000	0.3099	0.3088	0.3111	1.8221	1.8224	1.8209
0.7000	0.3199	0.3197	0.3205	2.0138	2.0158	2.0183
0.8000	0.3223	0.3225	0.3234	2.2255	2.2246	2.2217
0.9000	0.3185	0.3185	0.3165	2.4596	2.4594	2.4610
1.0000	0.3096	0.3093	0.3077	2.7183	2.7149	2.7205
1.1000	0.2967	0.2960	0.2969	3.0042	3.0043	3.0031
1.2000	0.2807	0.2802	0.2816	3.3201	3.3197	3.3211
1.3000	0.2626	0.2632	0.2644	3.6693	3.6693	3.6704
1.4000	0.2430	0.2431	0.2458	4.0552	4.0549	4.0535
1.5000	0.2226	0.2229	0.2213	4.4817	4.4819	4.4822
1.6000	0.2018	0.2017	0.2022	4.9530	4.9561	4.9557
1.7000	0.1812	0.1818	0.1789	5.4739	5.4740	5.4781
1.8000	0.1610	0.1619	0.1605	6.0496	6.0500	6.0510
1.9000	0.1415	0.1416	0.1421	6.6859	6.6900	6.6823
2.0000	0.1231	0.1230	0.1226	7.3891	7.3889	7.3857

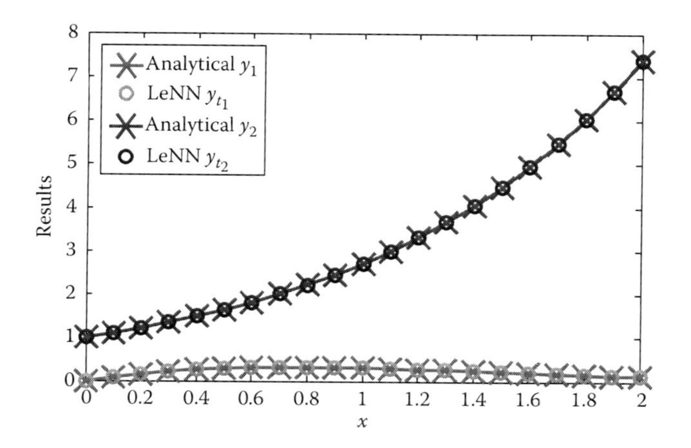

FIGURE 5.6
Plot of analytical and LeNN results (Example 5.3).

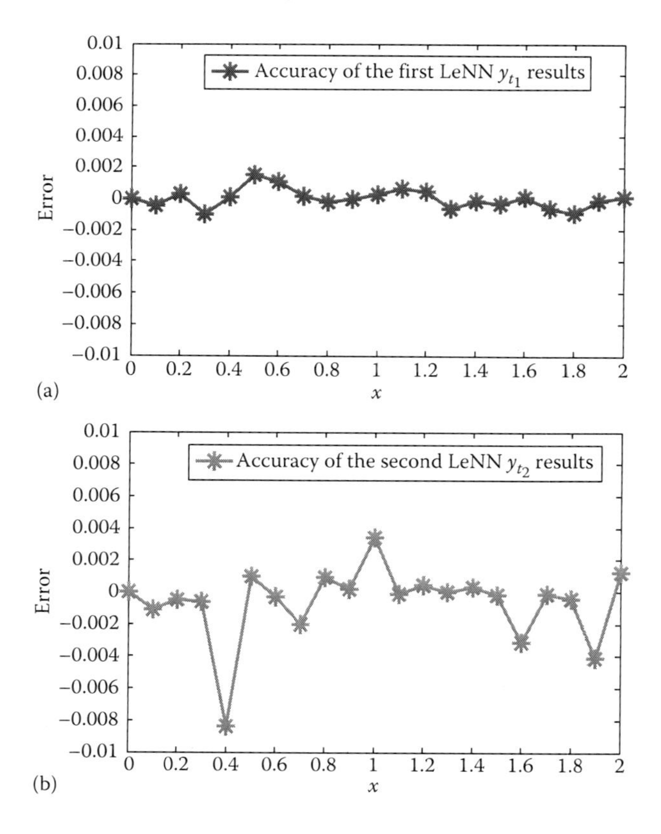

FIGURE 5.7
Error plot between analytical and LeNN results (Example 5.3). (a) Error plot of y_{t_1}. (b) Error plot of y_{t_2}.

TABLE 5.4

Analytical and LeNN Results for Testing Points (Example 5.3)

Testing Points	0.3894	0.7120	0.9030	1.2682	1.5870	1.8971
Analytical y_1	0.2572	0.3206	0.3183	0.2686	0.2045	0.1421
LeNN y_{t_1}	0.2569	0.3210	0.3180	0.2689	0.2045	0.1420
Analytical y_2	1.4761	2.0381	2.4670	3.5544	4.8891	6.6665
LeNN y_{t_2}	1.4760	2.0401	2.4672	3.5542	4.8894	6.6661

References

1. S. Mall and S. Chakraverty. Chebyshev neural network based model for solving Lane–Emden type equations. *Applied Mathematics and Computation*, 247: 100–114, November 2014.
2. S. Mall and S. Chakraverty. Numerical solution of nonlinear singular initial value problems of Emden–Fowler type using Chebyshev neural network method. *Neurocomputing*, 149: 975–982, February 2015.
3. S. Mall and S. Chakraverty. Application of Legendre neural network for solving ordinary differential equations, *Applied Soft Computing*, 43: 347–356, 2016.
4. S. Mall and S. Chakraverty. Multi layer versus functional link single layer neural network for solving nonlinear singular initial value problems. *Third International Symposium on Women computing and Informatics (WCI-2015)*, SCMS College, Kochi, Kerala, India, August 10–13, Published in Association for Computing Machinery (ACM) Proceedings, pp. 678–683, 2015.
5. S. Mall and S. Chakraverty. Hermite functional link neural network for solving the Van der Pol-Duffing oscillator equation. *Neural Computation*, 28(8): 1574–1598, July 2016.
6. Y.H. Pao and S.M. Philips. The functional link net and learning optimal control. *Neurocomputing*, 9(2): 149–164, October 1995.
7. D.R. Parisi, M.C. Mariani, and M.A. Laborde. Solving differential equations with unsupervised neural networks. *Chemical Engineering and Processing*, 42: 715–721, September 2003.
8. S. Purwar, I.N. Kar, and A.N. Jha. Online system identification of complex systems using Chebyshev neural network. *Applied Soft Computing*, 7(1): 364–372, January 2007.
9. J.C. Patra, R.N. Pal, B.N. Chatterji, and G. Panda. Identification of nonlinear dynamic systems using functional link artificial neural networks. *IEEE Transactions on Systems, Man, and Cybernetics, Part B—Cybernetics*, 29(2): 254–262, April 1999.
10. J.C. Patra and A.C. Kot. Nonlinear dynamic system identification using Chebyshev functional link artificial neural network. *IEEE Transactions on Systems, Man and Cybernetics, Part B—Cybernetics*, 32(4): 505–511, August 2002.
11. J.C. Patra, M. Juhola, and P.K. Meher. Intelligent sensors using computationally efficient Chebyshev neural networks. *IET Science, Measurement and Technology*, 2(2): 68–75, March 2008.

12. W.-D. Weng, C.-S. Yang, and R.-C. Lin. A channel equalizer using reduced decision feedback Chebyshev functional link artificial neural networks. *Information Sciences*, 177(13): 2642–2654, July 2007.
13. T.T. Lee and J.T. Jeng. The Chebyshev-polynomials-based unified model neural networks for function approximation. *IEEE Transactions on Systems, Man and Cybernetics, Part B—Cybernetics*, 28(6): 925–935, December 1998.
14. S.S. Yang and C.S. Tseng. An orthogonal neural network for function approximation. *IEEE Transactions on Systems, Man, and Cybernetics, Part B—Cybernetics*, 26(5): 779–784, October 1996.
15. J.C. Patra, W.C. Chin, P.K. Meher, and G. Chakraborty. Legendre-FLANN-based nonlinear channel equalization in wireless. *IEEE International Conference on Systems, Man and Cybernetics*, Singapore, pp. 1826–1831, October 2008.
16. J.C. Patra, P.K. Meher, and G. Chakraborty. Nonlinear channel equalization for wireless communication systems using Legendre neural networks. *Signal Processing*, 89(11): 2251–2262, November 2009.
17. J.C. Patra and C. Bornand. Nonlinear dynamic system identification using Legendre neural network. *IEEE International Joint Conference on Neural Networks*, Barcelona, Spain, pp. 1–7, July 2010.
18. S.K. Nanda and D.P. Tripathy. Application of functional link artificial neural network for prediction of machinery noise in opencast mines. *Advances in Fuzzy Systems*, 2011: 1–11, April 2011.
19. R.B. Bhat and S. Chakraverty. *Numerical Analysis in Engineering*. Alpha Science Int. Ltd., Oxford, U.K., 2004.
20. C.F. Gerald and P.O. Wheatley. *Applied Numerical Analysis*. Pearson Education Inc., Boston, MA, 2004.
21. E. Suli and D.F. Mayers. *An Introduction to Numerical Analysis*. Cambridge University Press, Cambridge, U.K., 2003.
22. O.P. Singh, R.K. Pandey, and V.K. Singh. Analytical algorithm of Lane-Emden type equation arising in astrophysics using modified homotopy analysis method. *Computer Physics Communications*, 180(7): 1116–1124, July 2009.

6

Single-Layer Functional Link Artificial Neural Network with Regression-Based Weights

The objective of this chapter is to solve differential equations using the functional link artificial neural network (FLANN) model with regression-based weights. We introduce here a single-layer Chebyshev polynomial–based FLANN called Chebyshev neural network (ChNN) with regression-based weights to handle first- and higher-order ordinary differential equations (ODEs). In FLANN, the hidden layer is replaced by a functional expansion block for enhancement of the input patterns using Chebyshev polynomials. So the technique is computationally more efficient than the multilayer perceptron (MLP) network [1,2]. ChNN has been successfully applied in system identification [3,4], function approximation [5], digital communication [6], and solving differential equations [7–11], etc.

Here, we use the feed-forward neural network and error back-propagation method for minimizing the error function and modifying the parameters (weights). Initial weights from the input layer to the output layer are taken by a regression-based model. The number of parameters (weights) has been fixed according to the degree of the polynomial in the regression fitting. For this, the input and output data are fitted first with various-degree polynomials using a regression analysis and the coefficients involved are taken as initial weights to start with the neural training. Fixing of the weights depends upon the degree of the polynomial.

It is worth mentioning here that Chakraverty and his coauthors [12,13] investigated various application problems using the regression-based neural network (RBNN) model. Prediction of the response of structural systems subject to earthquake motions has been investigated by Chakraverty et al. [12] using the RBNN model. Chakraverty et al. [13] studied vibration frequencies of annular plates using RBNN. Recently, Mall and Chakraverty [14–17] developed the RBNN model for solving initial/boundary value problems of ODEs.

6.1 ChNN Model with Regression-Based Weights

This section incorporates the structure of the single-layer ChNN model with regression-based weights, its training algorithm, formulation, and gradient computation, respectively.

6.1.1 Structure of the ChNN Model

We consider here the single-layer ChNN model for the present investigation. Figure 6.1 shows the structure of the ChNN model, which consists of a single input node, a functional expansion block based on Chebyshev polynomials, and a single output node. Let us consider the single input node x having h number of data as $x=(x_1, x_2, \ldots, x_h)^T$. It may be noted that Chebyshev polynomials are a set of orthogonal polynomials, and the first two Chebyshev polynomials may be written as

$$\left.\begin{aligned} T_0(x) &= 1 \\ T_1(x) &= x \end{aligned}\right\} \quad (6.1)$$

Higher-order Chebyshev polynomials may be generated by the well-known recursive formula

$$T_{r+1}(x) = 2xT_r(x) - T_{r-1}(x) \quad (6.2)$$

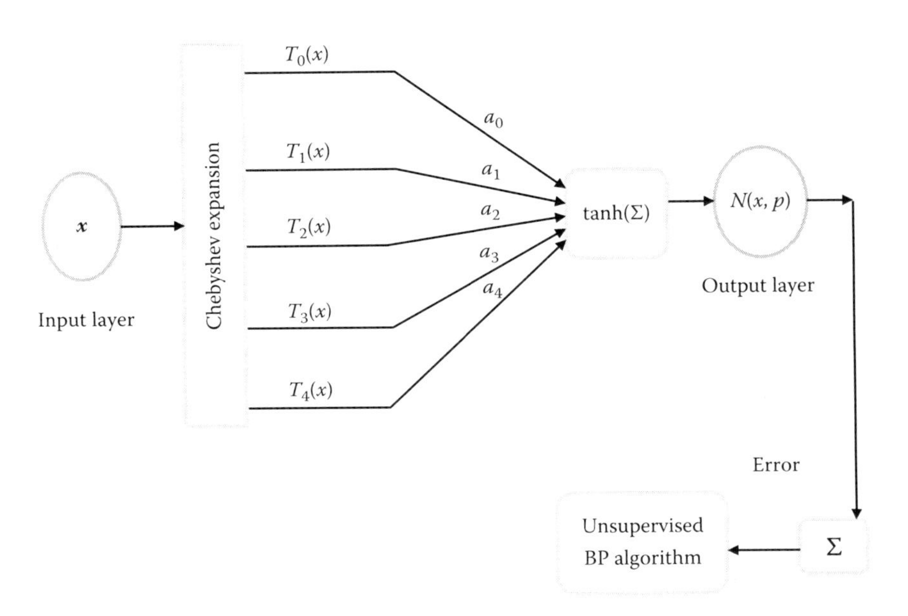

FIGURE 6.1
Structure of ChNN with regression-based weights.

The initial weights from the input layer to the output layer have been taken by regression fitting. If an nth-degree polynomial is considered, then the number of coefficients (constants) of the polynomial will be $n + 1$, and those may be considered as initial weights.

Let x and y be the input and output patterns. Then, polynomials of different degrees may be fitted. For example, if a polynomial of degree four is considered as [18,19]

$$y = p(x) = a_0 + a_1 x + a_2 x^2 + a_3 x^3 + a_4 x^4 \tag{6.3}$$

then the coefficients of the above polynomial, namely, a_0, a_1, a_2, a_3, and a_4, may be obtained by using the least-square fit method. As such, these constants may be taken now as the initial weights from the input layer to the output layer. Thus, depending upon the degree of the polynomial, the number of weights may be fixed.

6.1.2 Formulation and Gradient Computation of the ChNN Model

The previously discussed ChNN model is used here along with regression-based weights for solving ODEs. The difference here lies in the fixing of the initial weights only. In this regard, the structure, formulation, and error computation of ODEs using the ChNN model have already been discussed in Sections 5.1.1.1 and 5.1.1.2. The learning algorithm and gradient computation of the ChNN model are also discussed in Section 5.1.1.3. Figure 6.1 depicts the present model, where regression coefficients of a fourth-degree polynomial have been shown as the initial weights.

Now, we include in the next section a few example problems to demonstrate this method.

6.2 First-Order Linear ODEs

In this section, first-order ODEs have been considered to show the reliability of the method.

Example 6.1

An initial value problem (IVP) is considered as

$$\frac{dy}{dx} + 5y = e^{-3x}$$

subject to $y(0) = 0$.

TABLE 6.1

Comparison between Analytical and ChNN Results (with Regression-Based Weights) (Example 6.1)

Input Data	Analytical	ChNN with Regression-Based Weights
0	0	0
0.1	0.0671	0.0679
0.2	0.0905	0.0908
0.3	0.0917	0.0916
0.4	0.0829	0.0830
0.5	0.0705	0.0710
0.6	0.0578	0.0579
0.7	0.0461	0.0459
0.8	0.0362	0.0360
0.9	0.0280	0.0288
1.0	0.0215	0.0212

As discussed in Section 5.1.1.2, we can write the ChNN trial solution with regression-based weights as

$$y_t(x,p) = xN(x,p)$$

The network is trained for 10 equidistant points in [0, 1] and the first 5 Chebyshev polynomials and 5 regression-based weights from the input layer to the output layer. In Table 6.1, we compare analytical results with ChNN (with regression-based weights) results. Figure 6.2 plots this comparison between analytical and ChNN results. The error plot is depicted in Figure 6.3.

Example 6.2

Let us take another first-order ODE

$$\frac{dy}{dx} + 0.2y = e^{-0.2x}\cos x \quad x \in [0,1]$$

subject to $y(0) = 0$.

The ChNN trial solution with regression-based weights is formulated as

$$y_t(x,p) = xN(x,p)$$

Ten equidistant points in [0, 1] and five regression-based weights with respect to first five Chebyshev polynomials are considered. A comparison among

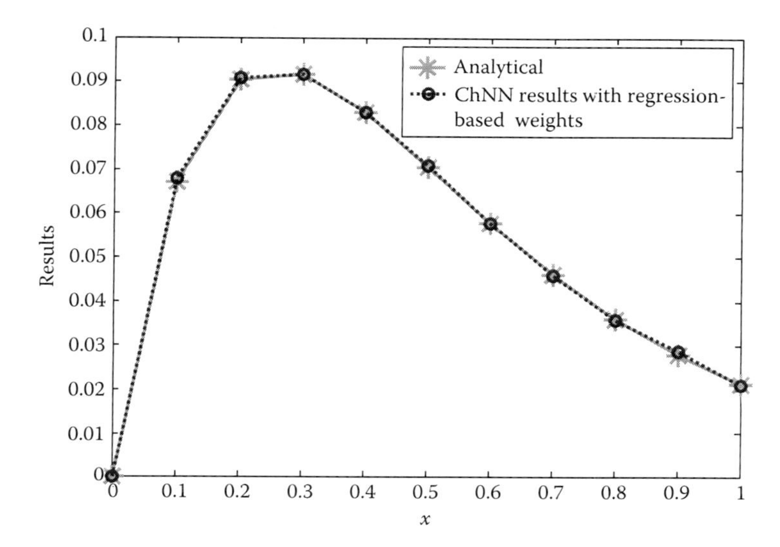

FIGURE 6.2
Plot of analytical and ChNN (with regression-based weights) results (Example 6.1).

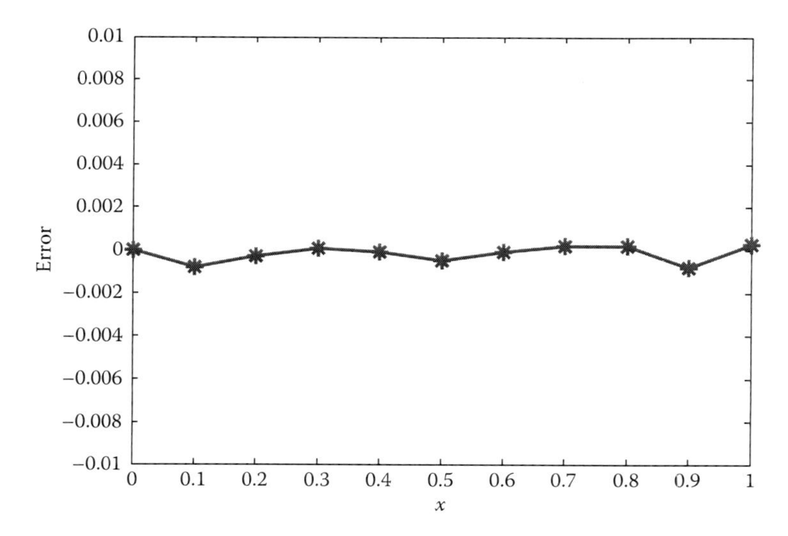

FIGURE 6.3
Error plot between analytical and ChNN (with regression-based weights) results (Example 6.1).

analytical results, ChNN results with regression-based weights, and ChNN results with arbitrary weights has been shown in Table 6.2. Figure 6.4 plots this comparison of analytical and ChNN results. The error (between analytical and ChNN results) is plotted in Figure 6.5.

It is worth mentioning that the CPU time for computation for the ChNN model with regression-based weights is 9,519.57 s, whereas the CPU time for the ChNN model with arbitrary weights is 15,847.38 s. As such, we can

TABLE 6.2

Comparison among Analytical Results, ChNN Results with Regression-Based Weights, and ChNN Results with Arbitrary Weights (Example 6.2)

Input Data	Analytical	ChNN with Regression-Based Weights	ChNN with Arbitrary Weights
0	0	0	0
0.1	0.0979	0.0979	0.0979
0.2	0.1909	0.1915	0.1912
0.3	0.2783	0.2786	0.2781
0.4	0.3595	0.3595	0.3598
0.5	0.4338	0.4335	0.4332
0.6	0.5008	0.5015	0.5009
0.7	0.5601	0.5620	0.5617
0.8	0.6113	0.6112	0.6110
0.9	0.6543	0.6549	0.6552
1	0.6889	0.6885	0.6880

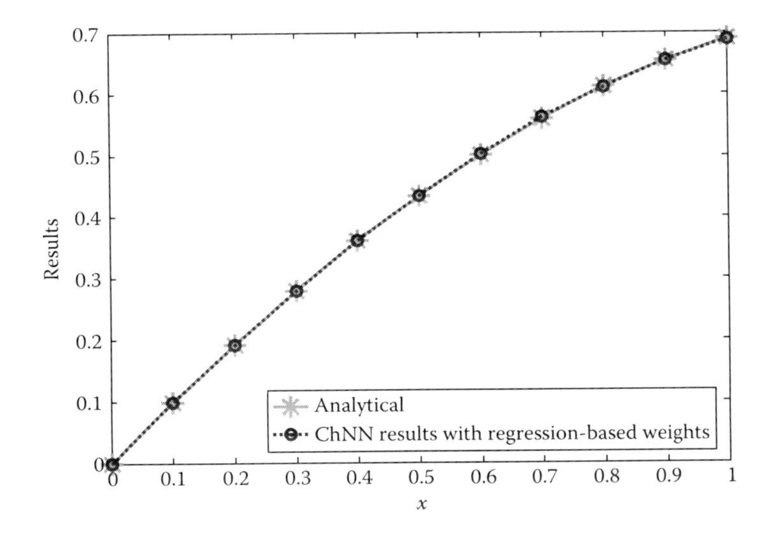

FIGURE 6.4
Plot of analytical and ChNN (with regression-based weights) results (Example 6.2).

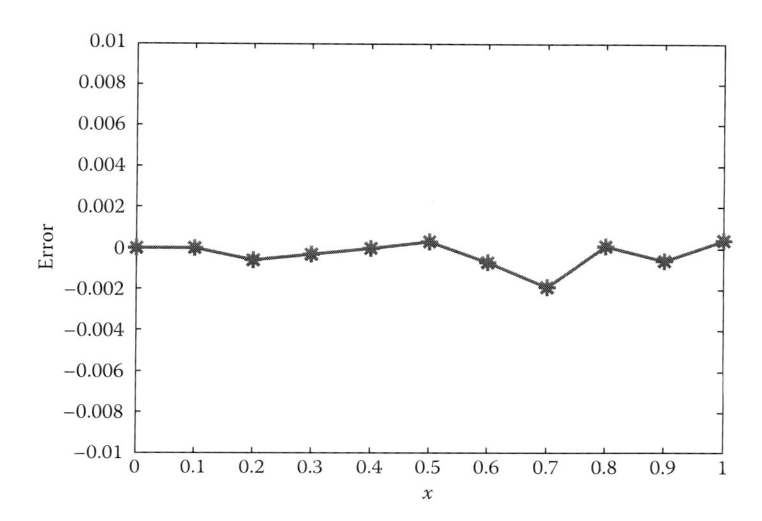

FIGURE 6.5
Error plot between analytical and ChNN (with regression-based weights) results (Example 6.2).

observe that ChNN with regression-based weights takes less CPU time for computation than ChNN with arbitrary weights.

6.3 Higher-Order ODEs

Here, a second-order IVP has been considered.

Example 6.3

In this example, we take a second-order ODE

$$2\frac{d^2y}{dx^2}+\frac{dy}{dx}=2x^2+3x+1$$

with initial conditions $y(0)=0$, $y'(0)=0$.

The ChNN trial solution with regression-based weights may be written as

$$y_t(x,p)=x^2N(x,p)$$

In this case also, the network is trained for 10 equidistant points in the time interval [0, 1]. We have fixed six weights according to regression fitting. A comparison of analytical and ChNN (with regression-based weights) results has been presented in Table 6.3. This comparison is also plotted in Figure 6.6.

TABLE 6.3

Comparison between Analytical and ChNN (with Regression-Based Weights) Results (Example 6.3)

Input Data	Analytical	ChNN with Regression-Based Weights
0	0	0
0.1	0.0028	0.0028
0.2	0.0121	0.0122
0.3	0.0298	0.0300
0.4	0.0578	0.0576
0.5	0.0981	0.0979
0.6	0.1529	0.1525
0.7	0.2245	0.2244
0.8	0.3155	0.3155
0.9	0.4282	0.4283
1	0.5655	0.5656

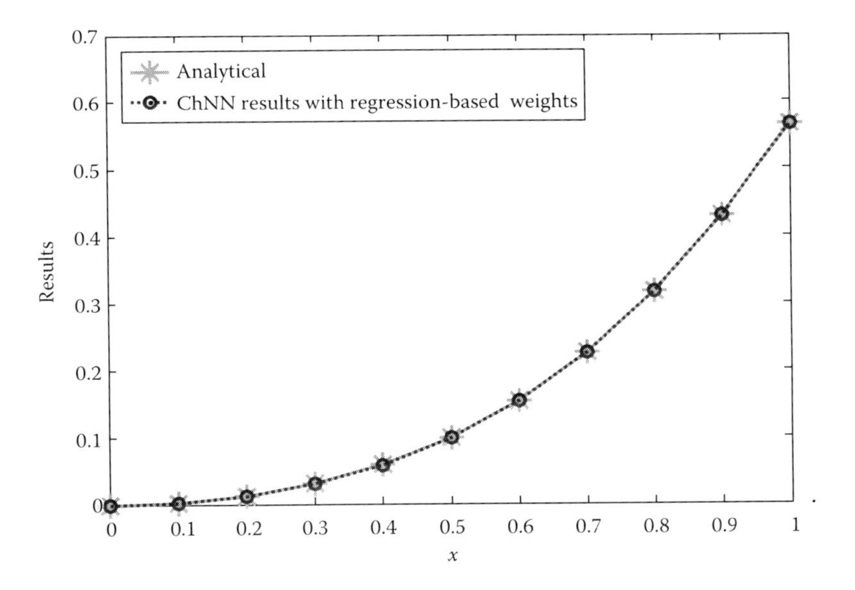

FIGURE 6.6
Plot of analytical and ChNN (with regression-based weights) results (Example 6.3).

References

1. A. Namatame and N. Ueda. Pattern classification with Chebyshev neural network. *International Journal Neural Network*, 3: 23–31, 1992.
2. J.C. Patra. Chebyshev neural network-based model for dual-junction solar cells. *IEEE Transactions on Energy Conversion*, 26: 132–140, 2011.
3. S. Purwar, I.N. Kar, and A.N. Jha. Online system identification of complex systems using Chebyshev neural network. *Applied Soft Computing*, 7: 364–372, 2007.
4. J.C. Patra and A.C. Kot. Nonlinear dynamic system identification using Chebyshev functional link artificial neural network. *IEEE Transactions on Systems, Man, and Cybernetics, Part B—Cybernetics*, 32(4): 505–511, 2002.
5. T.T. Lee and J.T. Jeng. The Chebyshev-polynomials-based unified model neural networks for function approximation. *IEEE Transactions on Systems, Man, and Cybernetics, Part B—Cybernetics*, 28(6): 925–935, 1998.
6. J.C. Patra, M. Juhola, and P.K. Meher. Intelligent sensors using computationally efficient Chebyshev neural networks. *IET Science, Measurement & Technology*, 2(2): 68–75, 2008.
7. S. Mall and S. Chakraverty. Chebyshev neural network based model for solving Lane–Emden type equations. *Applied Mathematics and Computation*, 247: 100–114, November 2014.
8. S. Mall and S. Chakraverty. Numerical solution of nonlinear singular initial value problems of Emden–Fowler type using Chebyshev neural network method. *Neurocomputing*, 149: 975–982, February 2015.
9. S. Mall and S. Chakraverty. Application of Legendre neural network for solving ordinary differential equations. *Applied Soft Computing*, 43: 347–356, 2016.
10. S. Mall and S. Chakraverty. Multi layer versus functional link single layer neural network for solving nonlinear singular initial value problems. *Third International Symposium on Women Computing and Informatics (WCI-2015)*, SCMS College, Kochi, Kerala, India, August 10–13, Published in Association for Computing Machinery (ACM) Proceedings, pp. 678–683, 2015.
11. S. Mall and S. Chakraverty. Hermite functional link neural network for solving the Van der Pol-Duffing oscillator equation. *Neural Computation*, 28(8): 1574–1598, July 2016.
12. S. Chakraverty, V.P. Singh, and R.K. Sharma. Regression based weight generation algorithm in neural network for estimation of frequencies of vibrating plates. *Journal of Computer Methods in Applied Mechanics and Engineering*, 195: 4194–4202, July 2006.
13. S. Chakraverty, V.P. Singh, R.K. Sharma, and G.K. Sharma. Modelling vibration frequencies of annular plates by regression based neural Network. *Applied Soft Computing*, 9(1): 439–447, January 2009.
14. S. Mall and S. Chakraverty. Comparison of artificial neural network architecture in solving ordinary differential equations. *Advances in Artificial Neural Systems*, 2013: 1–24, October 2013.

15. S. Mall and S. Chakraverty. Regression-based neural network training for the solution of ordinary differential equations. *International Journal of Mathematical Modelling and Numerical Optimization*, 4(2): 136–149, 2013.
16. S. Chakraverty and S. Mall. Regression based weight generation algorithm in neural network for solution of initial and boundary value problems. *Neural Computing and Applications*, 25(3): 585–594, September 2014.
17. S. Mall and S. Chakraverty. Regression based neural network model for the solution of initial value problem. *National Conference on Computational and Applied Mathematics in Science and Engineering (CAMSE-2012)*, VNIT, Nagpur, India, December 2012.
18. S.T. Karris. *Numerical Analysis: Using MATLAB and Spreadsheets*. Orchard Publication, Fremont, CA, 2004.
19. R.B. Bhat and S. Chakraverty. *Numerical Analysis in Engineering*. Alpha Science Int. Ltd., Oxford, U.K., 2004.

7

Lane–Emden Equations

Many problems in astrophysics and mathematical physics can be modeled by initial or boundary value problems, namely, as second-order nonlinear ordinary differential equations (ODEs). In astrophysics, the equation that describes the equilibrium density distribution in a self-gravitating sphere of polytrophic isothermal gas was proposed by Lane [1] and further described by Emden [2], which is now known as the Lane–Emden equation. The general form of the Lane–Emden equation is as follows:

$$\frac{d^2y}{dx^2}+\frac{2}{x}\frac{dy}{dx}+f(x,y)=g(x) \quad x\geq 0$$
$$\text{with initial conditions } y(0)=\alpha, y'(0)=0 \tag{7.1}$$

where

$f(x,y)$ is a nonlinear function of x and y

$g(x)$is a function of x

These Lane–Emden-type equations are singular at x = 0. So the analytical solution of this type of equation is possible in the neighborhood of the singular point [3]. In Equation 7.1, $f(x,y)$ describes several phenomena in astrophysics such as the theory of stellar structure, the thermal behavior of a spherical cloud of gas, isothermal gas spheres, etc. The most popular form of $f(x,y)$ is as follows:

$$f(x,y)=y^m, \quad y(0)=1, \quad y'(0)=0, \quad \text{and} \quad g(x)=0 \tag{7.2}$$

So the standard form of the Lane–Emden equation can be written as

$$\frac{d^2y}{dx^2}+\frac{2}{x}\frac{dy}{dx}+y^m=0 \tag{7.3}$$

$$\Rightarrow \frac{1}{x^2}\frac{d}{dx}\left(x^2\frac{dy}{dx}\right)+y^m=0$$
$$\text{with initial conditions } y(0)=1, y'(0)=0. \tag{7.4}$$

The Lane–Emden equation is a dimensionless form of Poisson's equation for the gravitational potential of a Newtonian self-gravitating, spherically symmetric, polytrophic fluid. Here, m is a constant, which is called the polytrophic index. Equation 7.3 describes the thermal behavior of a spherical cloud of gas acting under the mutual attraction of its molecules and subject to the classical laws of thermodynamics. Another nonlinear form of $f(x,y)$ is the exponential function, that is,

$$f(x,y) = e^{y} \tag{7.5}$$

Substituting (7.5) in Equation 7.3, we have

$$\frac{d^2y}{dx^2} + \frac{2}{x}\frac{dy}{dx} + e^{y} = 0 \tag{7.6}$$

This equation describes the behavior of isothermal gas spheres where the temperature remains constant.

Substituting $f(x,y) = e^{-y}$ in Equation 7.3

$$\frac{d^2y}{dx^2} + \frac{2}{x}\frac{dy}{dx} + e^{-y} = 0 \tag{7.7}$$

which gives a model that appears in the theory of thermionic current and has thoroughly been investigated by Richardson [4].

Exact solutions of Equation 7.3 for $m = 0$, 1, and 5 have been obtained by Chandrasekhar [5] and Datta [6]. For $m = 5$, only one parameter family of solution is obtained in [7]. For other values of m, the standard Lane–Emden equations can only be solved numerically. The singularity behavior at the origin, that is, at $x = 0$, gives rise to a challenge to the solution of not only the Lane–Emden equations but also various other linear and nonlinear initial value problems (IVPs) in astrophysics and quantum mechanics. Different analytical approaches based on either series solutions or perturbation techniques have been used to handle the Lane–Emden equations. In this regard, the Adomian decomposition method (ADM) has been used by Wazwaz [8–10] and Shawagfeh [11] to handle the Lane–Emden equations. Chowdhury et al. [12,13] employed the homotopy perturbation method (HPM) to solve singular IVPs of time-independent equations. Liao [14] presented an algorithm based on ADM for solving Lane–Emden-type equations. An approximate solution of a differential equation arising in astrophysics using the variational iteration method has been developed by Dehghan and Shakeri [15]. The Emden–Fowler equation has been solved by utilizing the techniques of Lie and Painleve analysis proposed by Govinder and Leach [16].

An efficient analytical algorithm based on the modified homotopy analysis method has been presented by Singh et al. [17]. Muatjetjeja and Khalique in [18] provided an exact solution of the generalized Lane–Emden equations of the first and second kind. Recently, Mall and Chakraverty [19] have developed ChNN model for solving second-order singular initial value problems of Lane–Emden type.

The aim of the present chapter is to use multilayer ANN and single-layer functional link artificial neural network (FLANN) models for solving homogeneous and nonhomogeneous Lane–Emden equations. Various ANN models are used here to overcome singularity. We have used the unsupervised version of error back-propagation algorithm for minimizing the error function and updating the network parameters. Initial weights from the input layer to the output layer are considered as random.

7.1 Multilayer ANN-Based Solution of Lane–Emden Equations

The formulation, error computation, and learning algorithm for a multilayer ANN are already discussed in Section 3.2.2.2 (Equations 3.16 through 3.22). As such, here, we have considered two example problems using a multilayer ANN.

Example 7.1

In this example, we take a homogeneous Lane–Emden equation with $f(x,y)=y^5$

$$\frac{d^2y}{dx^2}+\frac{2}{x}\frac{dy}{dx}+y^5=0 \quad x\geq 0$$

with initial conditions $y(0)=1$, $y'(0)=0$.

The ANN trial solution for this problem, as given in Section 3.2.2.2 (Equation 3.17), can be expressed as

$$y_t(x,p)=1+x^2N(x,p)$$

Here, we have trained the network for 20 equidistant points in [0, 1]. A comparison between analytical and multilayer ANN results is shown in Table 7.1. Figure 7.1 depicts this comparison of results. The error plot is shown in Figure 7.2.

TABLE 7.1

Comparison between Analytical and ANN Results (Example 7.1)

Input Data	Analytical [3]	ANN	Absolute Error
0	1.0000	1.0001	0.0001
0.1	0.9983	0.9972	0.0011
0.15	0.9963	0.9965	0.0002
0.2	0.9934	0.9936	0.0002
0.25	0.9897	0.9891	0.0006
0.3	0.9853	0.9890	0.0037
0.35	0.9802	0.9816	0.0014
0.4	0.9744	0.9742	0.0002
0.45	0.9679	0.9658	0.0021
0.5	0.9608	0.9572	0.0036
0.55	0.9531	0.9539	0.0008
0.6	0.9449	0.9467	0.0018
0.65	0.9362	0.9355	0.0007
0.7	0.9271	0.9288	0.0017
0.75	0.9177	0.9173	0.0004
0.8	0.9078	0.9061	0.0017
0.85	0.8977	0.8933	0.0044
0.9	0.8874	0.8836	0.0038
0.95	0.8768	0.8768	0
1.0	0.8660	0.8680	0.0020

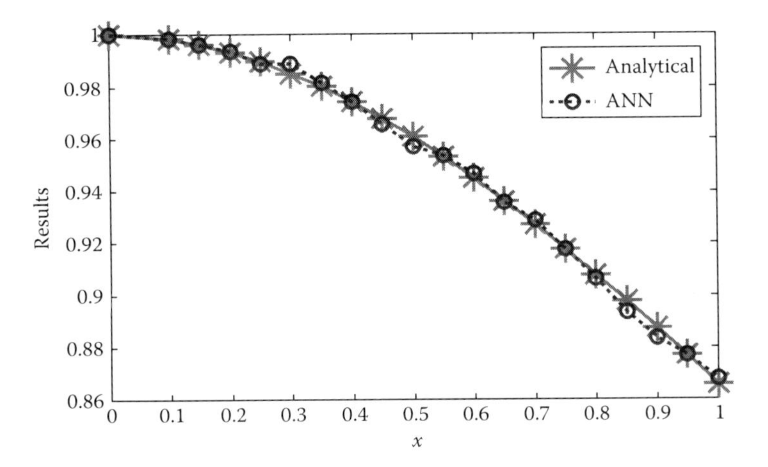

FIGURE 7.1
Plot of analytical and ANN results (Example 7.1).

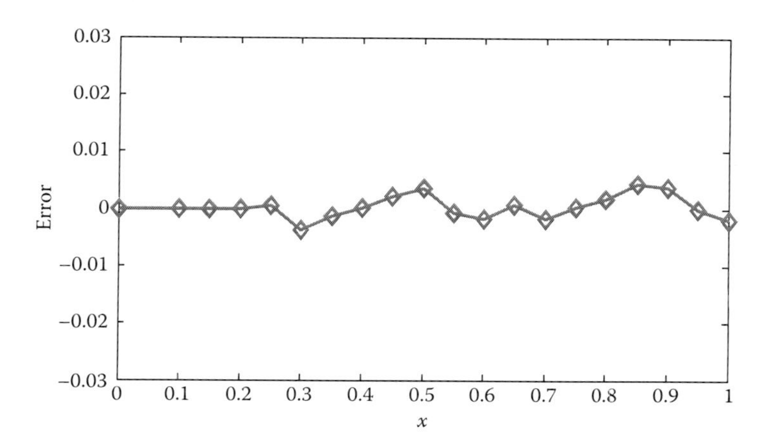

FIGURE 7.2
Error plot between analytical and ANN results (Example 7.1).

Example 7.2

A Lane–Emden equation with $f(x,y)=x^m e^{-y}$ is considered here:

$$\frac{d^2y}{dx^2}+\frac{2}{x}\frac{dy}{dx}+x^m e^{-y}=0$$

with initial conditions $y(0)=0$, $y'(0)=0$.

For $m = 0$, this equation models Richardson's theory of thermionic current when the density and electric force of an electron gas in the neighborhood of a hot body in thermal equilibrium is to be determined [4]. A particular solution by ADM of this problem is given in [10]

$$y(x)=\ln\left(-\frac{x^2}{2}\right)$$

The ANN trial solution is written as

$$y_t(x,p)=x^2N(x,p)$$

In this case, 10 equidistant points in [0, 1] are considered. Table 7.2 shows ADM and ANN results. A comparison of these ADM (particular) and ANN results is presented in Figure 7.3. Lastly, Figure 7.4 depicts the plot of error between ADM and ANN results at some testing points are given in Table 7.3.

TABLE 7.2

Comparison between ADM and ANN Results (Example 7.2)

Input Data	ADM [10]	ANN
0	0.0000	0.0000
0.1	−5.2983	−5.2916
0.2	−3.9120	−3.9126
0.3	−3.1011	−3.1014
0.4	−2.5257	−2.5248
0.5	−2.0794	−2.0793
0.6	−1.7148	−1.7159
0.7	−1.4065	−1.4078
0.8	−1.1394	−1.1469
0.9	−0.9039	−0.9048
1.0	−0.6931	−0.7001

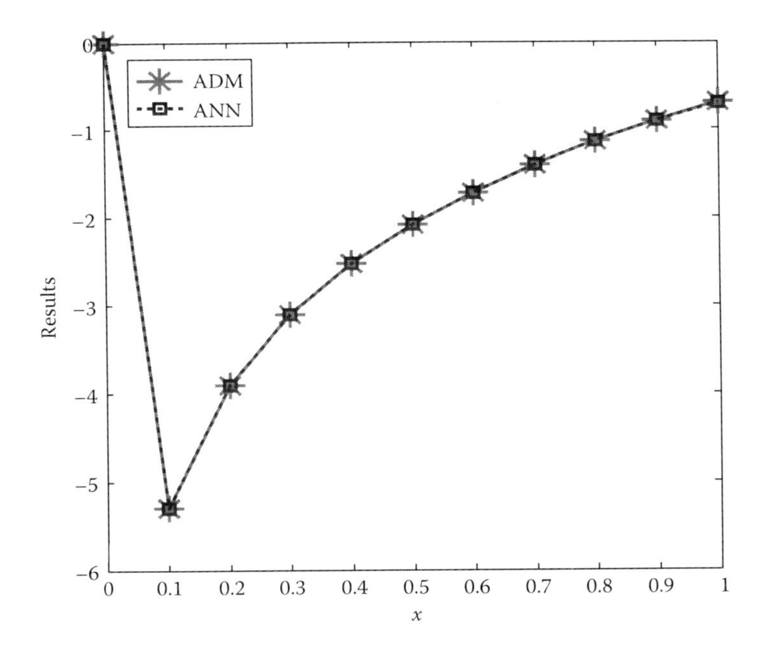

FIGURE 7.3
Plot of ADM and ANN results (Example 7.2).

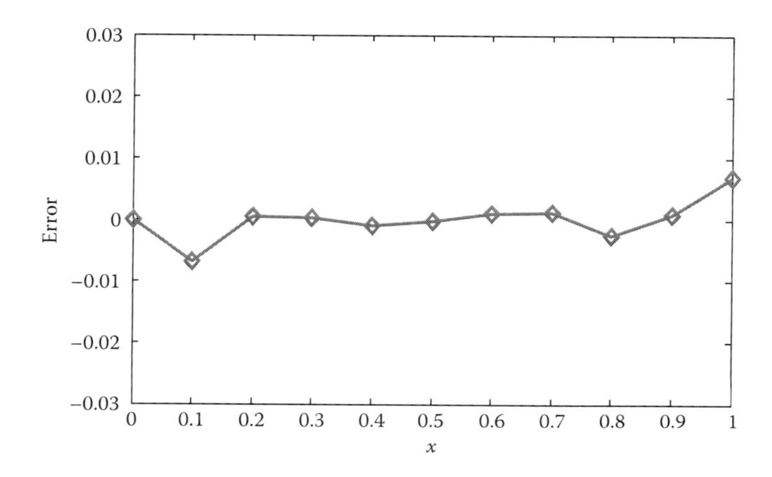

FIGURE 7.4
Error plot between ADM and ANN results (Example 7.2).

TABLE 7.3
ADM and ANN Results for Testing Points (Example 7.2)

Testing Points	0.209	0.399	0.513	0.684	0.934
Particular	−3.8239	−2.5307	−2.0281	−1.4527	−0.8297
ANN	−3.8236	−2.5305	−2.0302	−1.4531	−0.8300

7.2 FLANN-Based Solution of Lane–Emden Equations

In this section, a single-layer FLANN model has been used to solve second-order singular nonlinear ODEs of Lane–Emden equations.

The Chebyshev neural network (ChNN) model has been used for solving Lane–Emden equations. In this regard, the structure, formulation, and error computation of ODEs using the ChNN model have already been discussed in Sections 5.1.1.1 and 5.1.1.2. The ChNN trial solution for a second-order IVP is the same as the trial solution of a multilayer ANN. The learning algorithm and gradient computation for ChNN are discussed in Section 5.1.1.3.

Here, homogeneous and nonhomogeneous Lane–Emden equations have been considered to show the reliability of the ChNN model. Here, we have trained the ChNN model for different numbers of training points (such as 10, 15, 20, etc.) because various problems converge with different numbers of training points. In each problem, the number of points taken is mentioned, which gives a good result with acceptable accuracy.

7.2.1 Homogeneous Lane–Emden Equations

It was physically shown in [6,7] that m can have the values in the interval [0, 5] and exact solutions exist only for m = 0, 1, and 5. So we have computed the ChNN solution with these particular values of m, and these will be compared with the known exact solutions to gain confidence in our present methodology.

The standard Lane–Emden equations are discussed in Examples 7.3 through 7.6 for index values m = 0, 1, 0.5, and 2.5, respectively.

Example 7.3

For $m = 0$, the equation becomes a linear ODE

$$\frac{d^2y}{dx^2} + \frac{2}{x}\frac{dy}{dx} + 1 = 0$$

with initial conditions $y(0) = 1,\ y'(0) = 0$.

As discussed in Section 5.1.1.2, we can write the ChNN trial solution as

$$y_t(x,p) = 1 + x^2 N(x,p)$$

The network is trained for 10 equidistant points in [0, 1] and the first 5 Chebyshev polynomials and 5 weights from the input layer to the output layer. In Table 7.4, we compare analytical solutions with ChNN solutions with arbitrary weights. Figure 7.5 plots this comparison between analytical and ChNN results. The error plot is depicted in Figure 7.6.

TABLE 7.4

Comparison between Analytical and ChNN Results (Example 7.3)

Input Data	Analytical [3]	ChNN	Relative Error
0	1.0000	1.0000	0
0.1	0.9983	0.9993	0.0010
0.2	0.9933	0.9901	0.0032
0.3	0.9850	0.9822	0.0028
0.4	0.9733	0.9766	0.0033
0.5	0.9583	0.9602	0.0019
0.6	0.9400	0.9454	0.0054
0.7	0.9183	0.9139	0.0044
0.8	0.8933	0.8892	0.0041
0.9	0.8650	0.8633	0.0017
1.0	0.8333	0.8322	0.0011

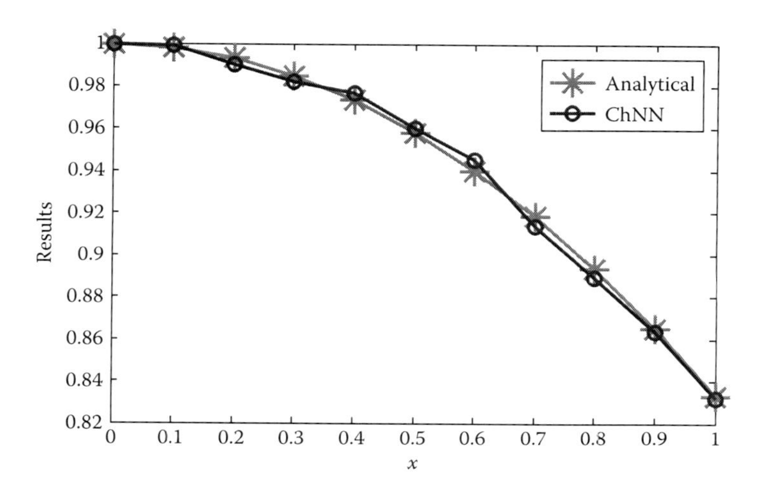

FIGURE 7.5
Plot of analytical and ChNN results (Example 7.3).

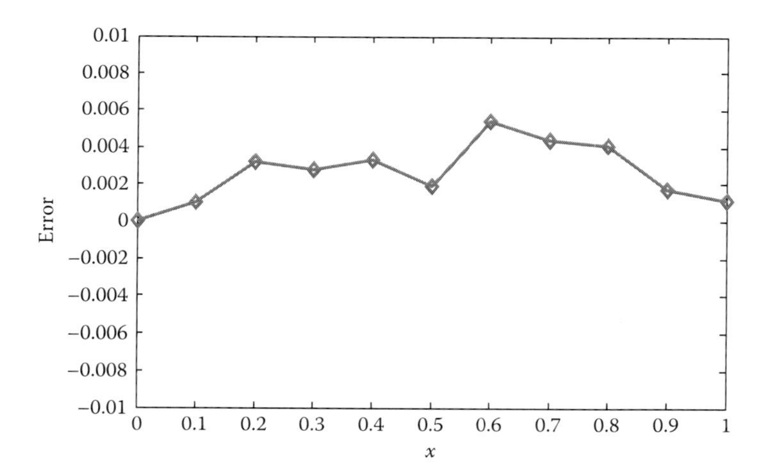

FIGURE 7.6
Error plot of analytical and ChNN results (Example 7.3).

Example 7.4

Let us consider a Lane–Emden equation for $m = 1$ with the same initial conditions

$$\frac{d^2y}{dx^2} + \frac{2}{x}\frac{dy}{dx} + y = 0$$

The ChNN trial solution, in this case, is the same as in Example 7.3.

TABLE 7.5

Comparison between Analytical and ChNN Results (Example 7.4)

Input Data	Analytical [3]	ChNN	Relative Error
0	1.0000	1.0000	0
0.1000	0.9983	1.0018	0.0035
0.1500	0.9963	0.9975	0.0012
0.2000	0.9933	0.9905	0.0028
0.2500	0.9896	0.9884	0.0012
0.3000	0.9851	0.9839	0.0012
0.3500	0.9797	0.9766	0.0031
0.4000	0.9735	0.9734	0.0001
0.4500	0.9666	0.9631	0.0035
0.5000	0.9589	0.9598	0.0009
0.5500	0.9503	0.9512	0.0009
0.6000	0.9411	0.9417	0.0006
0.6500	0.9311	0.9320	0.0009
0.7000	0.9203	0.9210	0.0007
0.7500	0.9089	0.9025	0.0064
0.8000	0.8967	0.8925	0.0042
0.8500	0.8839	0.8782	0.0057
0.9000	0.8704	0.8700	0.0004
0.9500	0.8562	0.8588	0.0026
1.0000	0.8415	0.8431	0.0016

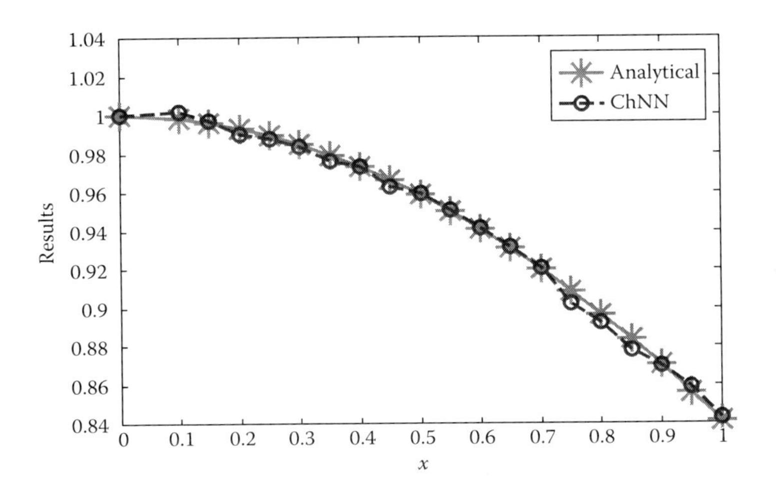

FIGURE 7.7
Plot of analytical and ChNN results (Example 7.4).

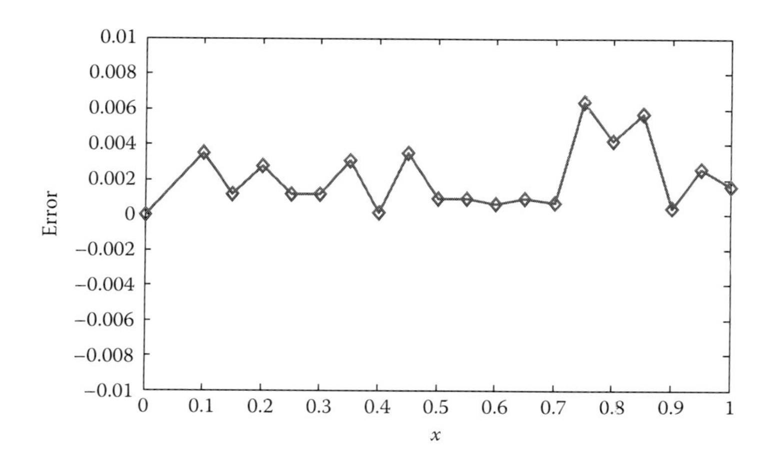

FIGURE 7.8
Error plot of analytical and ChNN results (Example 7.4).

Twenty equidistant points in [0, 1] and five weights with respect to the first five Chebyshev polynomials are considered. A comparison of analytical and ChNN results has been shown in Table 7.5. Figure 7.7 depicts this comparison. Figure 7.8 shows the plot of error between analytical and ChNN results.

From the table, one can observe that the exact (analytical) results compare very well with the ChNN results. As such, next, we have given some examples with values of $m = 0.5$ and 2.5 to get new approximate results of the said differential equation.

Example 7.5

Let us consider a Lane–Emden equation for $m = 0.5$

$$\frac{d^2y}{dx^2} + \frac{2}{x}\frac{dy}{dx} + y^{0.5} = 0$$

with initial conditions $y(0) = 1,\ y'(0) = 0$.

The ChNN trial solution is written as

$$y_t(x,p) = 1 + x^2 N(x,p)$$

Ten equidistant points and five weights with respect to the first five Chebyshev polynomials are considered here to train the model. Table 7.6 includes ChNN and HPM [13] results along with the relative error at the given points. A plot of error (between ChNN and HPM results) is presented in Figure 7.9.

TABLE 7.6
Comparison between ChNN and HPM Results (Example 7.5)

Input Data	ChNN	HPM [13]	Relative Error
0	1.0000	1.0000	0
0.1	0.9968	0.9983	0.0015
0.2	0.9903	0.9933	0.0030
0.3	0.9855	0.9850	0.0005
0.4	0.9745	0.9734	0.0011
0.5	0.9598	0.9586	0.0012
0.6	0.9505	0.9405	0.0100
0.7	0.8940	0.9193	0.0253
0.8	0.8813	0.8950	0.0137
0.9	0.8597	0.8677	0.0080
1.0	0.8406	0.8375	0.0031

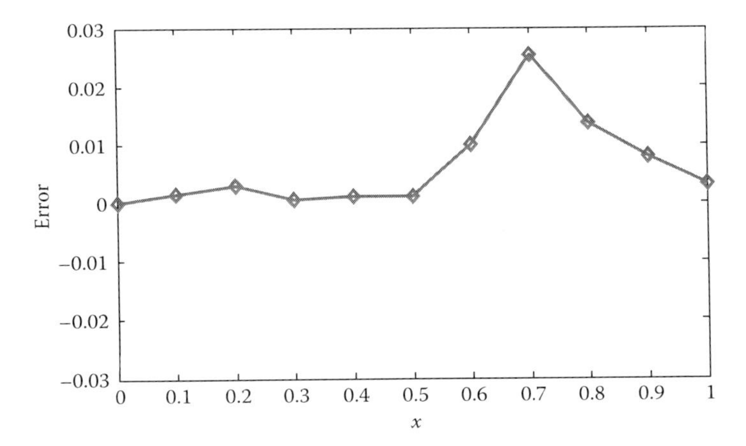

FIGURE 7.9
Error plot between ChNN and HPM results (Example 7.5).

Example 7.6

Here, we take a Lane–Emden equation for $m = 2.5$ with the same initial conditions as

$$\frac{d^2y}{dx^2} + \frac{2}{x}\frac{dy}{dx} + y^{2.5} = 0$$

The ChNN trial solution is the same as in Example 7.5.

TABLE 7.7

Comparison between ChNN and HPM Results (Example 7.6)

Input Data	ChNN	HPM [13]	Relative Error
0	1.0000	1.0000	0
0.1	0.9964	0.9983	0.0019
0.2	0.9930	0.9934	0.0004
0.3	0.9828	0.9852	0.0024
0.4	0.9727	0.9739	0.0012
0.5	0.9506	0.9596	0.0090
0.6	0.9318	0.9427	0.0109
0.7	0.9064	0.9233	0.0169
0.8	0.8823	0.9019	0.0196
0.9	0.8697	0.8787	0.0090
1.0	0.8342	0.8542	0.0200

Again, 10 points in the given domain and 5 weights are considered to train the ChNN model. Table 7.7 shows ChNN and HPM [13] results along with the relative error.

Example 7.7

Here, we now consider an example of the Lane–Emden equation with $f(x,y) = -2(2x^2+3)y$.

As such, a second-order homogeneous Lane–Emden equation will be

$$\frac{d^2y}{dx^2} + \frac{2}{x}\frac{dy}{dx} - 2\left(2x^2+3\right)y = 0 \quad x \geq 0$$

with initial conditions $y(0)=1$, $y'(0)=0$.

As discussed in Section 5.1.1.2, we can write the ChNN trial solution as

$$y_t\left(x,p\right) = 1 + x^2 N\left(x,p\right)$$

We have trained the network for 10 equidistant points in [0, 1]. As in the previous case, analytical and obtained ChNN results are shown in Table 7.8. ChNN results at testing points are given in Table 7.9. Lastly, the error (between analytical and ChNN results) is plotted in Figure 7.10.

TABLE 7.8

Comparison between Analytical and ChNN Results (Example 7.7)

Input Data	Analytical [17]	ChNN	Relative Error
0	1.0000	1.0000	0
0.1	1.0101	1.0094	0.0007
0.2	1.0408	1.0421	0.0013
0.3	1.0942	1.0945	0.0003
0.4	1.1732	1.1598	0.0134
0.5	1.2840	1.2866	0.0026
0.6	1.4333	1.4312	0.0021
0.7	1.6323	1.6238	0.0085
0.8	1.8965	1.8924	0.0041
0.9	2.2479	2.2392	0.0087
1.0	2.7148	2.7148	0

TABLE 7.9

ChNN Solutions for Testing Points (Example 7.7)

Testing Points	0.232	0.385	0.571	0.728	0.943
Analytical Results	1.0553	1.1598	1.3855	1.6989	2.4333
ChNN Results	1.0597	1.1572	1.3859	1.6950	2.4332

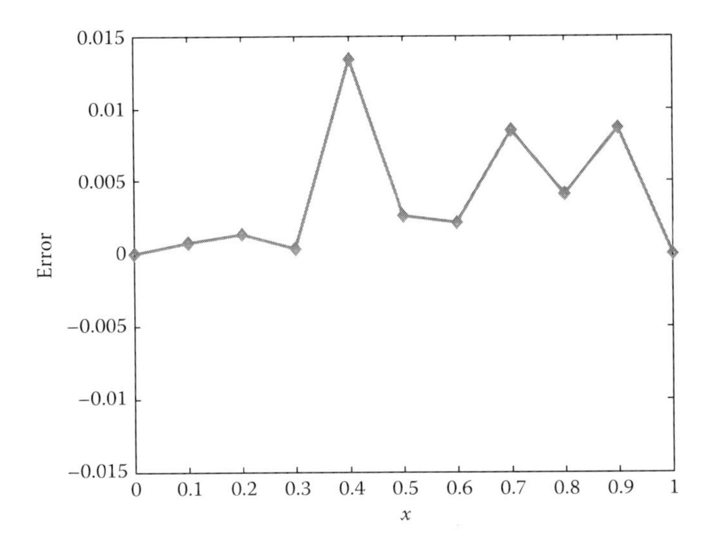

FIGURE 7.10
Error plot of analytical and ChNN results (Example 7.7).

7.2.2 Nonhomogeneous Lane–Emden Equation

In the following, nonhomogeneous Lane–Emden equation has been solved by [8,17] using ADM and the modified homotopy analysis method. Here, Example 7.8 is solved using the ChNN model.

Example 7.8

The nonhomogeneous Lane–Emden equation is written as

$$\frac{d^2y}{dx^2}+\frac{2}{x}\frac{dy}{dx}+y=6+12x+2x^2+x^3 \quad 0\le x\le 1$$

subject to $y(0)=0,\ y'(0)=0$.

This equation has the exact solution for $x\ge 0$ [17] as

$$y(x)=x^2+x^3$$

Here, the related ChNN trial solution is written as

$$y_t(x,p)=x^2N(x,p)$$

In this case, 20 equidistant points in [0, 1] and 5 weights with respect to the first 5 Chebyshev polynomials are considered. Table 7.10 shows analytical and

TABLE 7.10

Comparison between Analytical and ChNN Results (Example 7.8)

Input Data	Analytical [17]	ChNN	Relative Error
0	0	0	0
0.10	0.0110	0.0103	0.0007
0.15	0.0259	0.0219	0.0040
0.20	0.0480	0.0470	0.0010
0.25	0.0781	0.0780	0.0001
0.30	0.1170	0.1164	0.0006
0.35	0.1654	0.1598	0.0056
0.40	0.2240	0.2214	0.0026
0.45	0.2936	0.2947	0.0011
0.50	0.3750	0.3676	0.0074
0.55	0.4689	0.4696	0.0007
0.60	0.5760	0.5712	0.0048
0.65	0.6971	0.6947	0.0024
0.70	0.8330	0.8363	0.0033
0.75	0.9844	0.9850	0.0006
0.80	1.1520	1.1607	0.0087
0.85	1.3366	1.3392	0.0026
0.90	1.5390	1.5389	0.0001
0.95	1.7599	1.7606	0.0007
1.00	2.0000	2.0036	0.0036

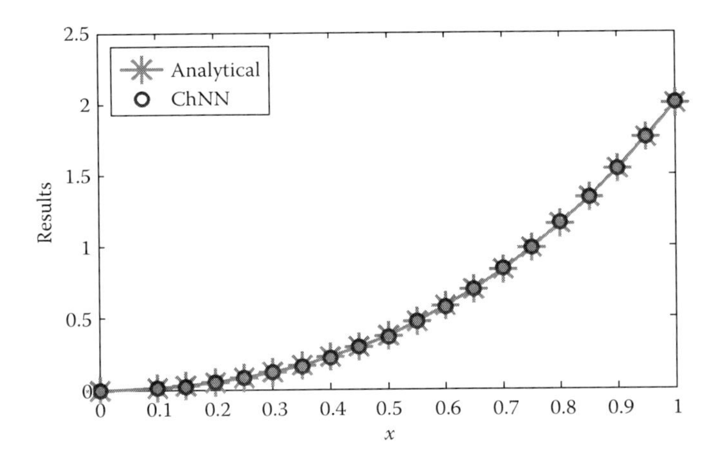

FIGURE 7.11
Plot of analytical and ChNN results (Example 7.8).

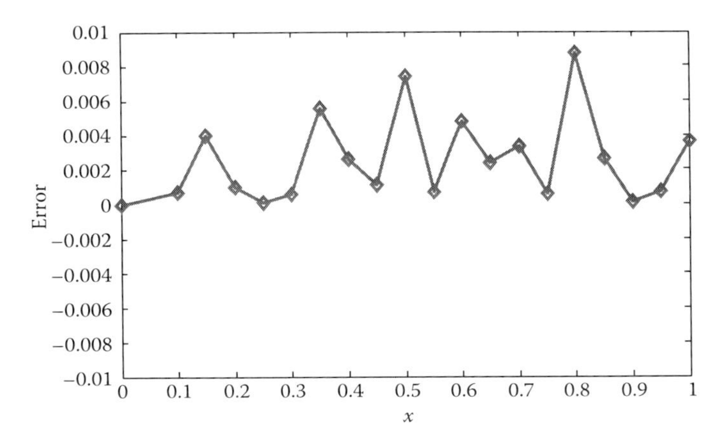

FIGURE 7.12
Error plot of analytical and ChNN results (Example 7.8).

ChNN results. These results are compared in Figure 7.11. Finally, Figure 7.12 depicts the plot of error between analytical and ChNN results.

References

1. J.H. Lane. On the theoretical temperature of the sun under the hypothesis of a gaseous mass maintaining its volume by its internal heat and depending on the laws of gases known to terrestrial experiment. *The American Journal of Science and Arts*, 50(2): 57–74, 1870.

2. R. Emden. *Gaskugeln, Anwendungen der mechanischenWarmen-theorie auf Kosmologie and meteorologische Problem*. Teubner, Leipzig, Germany, 1907.
3. H.T. Davis. *Introduction to Nonlinear Differential and Integral Equations*. Dover, New York, 1962.
4. O.W. Richardson. *The Emission of Electricity from Hot Bodies*. Longman, Green & Co., London, U.K., 1921.
5. S. Chandrasekhar. *Introduction to Study of Stellar Structure*. Dover, New York, 1967.
6. B.K. Datta. Analytic solution to the Lane-Emden equation. *Nuovo Cimento*, 111B: 1385–1388, 1996.
7. L. Dresner. *Similarity Solutions of Nonlinear Partial Differential Equations*. Pitman Advanced Publishing Program, London, U.K., 1983.
8. M. Wazwaz. A new algorithm for solving differential equation Lane–Emden type. *Applied Mathematics and Computation*, 118(3): 287–310, March 2001.
9. A.M. Wazwaz. Adomian decomposition method for a reliable treatment of the Emden–Fowler equation. *Applied Mathematics and Computation*, 161(2): 543–560, February 2005.
10. A.M. Wazwaz. The modified decomposition method for analytical treatment of differential equations. *Applied Mathematics and Computation*, 173(1): 165–176, February 2006.
11. N.T. Shawagfeh. Non perturbative approximate solution for Lane–Emden equation. *Journal of Mathematical Physics*, 34(9): 4364–4369, September 1993.
12. M.S.H. Chowdhury and I. Hashim. Solutions of a class of singular second order initial value problems by homotopy-perturbation Method. *Physics Letters A*, 365(5–6): 439–447, June 2007.
13. M.S.H. Chowdhury and I. Hashim. Solutions of Emden-Fowler Equations by homotopy-perturbation Method. *Nonlinear Analysis: Real Word Applications*, 10(1): 104–115, February 2009.
14. S.J. Liao. A new analytic algorithm of Lane–Emden type equations. *Applied Mathematics and Computation*, 142(1): 1–16, September 2003.
15. M. Dehghan and F. Shakeri. Approximate solution of a differential equation arising in astrophysics using the variational iteration method. *New Astronomy*, 13(1): 53–59, January 2008.
16. K.S. Govinder and P.G.L. Leach. Integrability analysis of the Emden-Fowler equation. *Journal of Nonlinear Mathematical Physics*, 14(3): 435–453, 2007.
17. O.P. Singh, R.K. Pandey, and V.K. Singh. Analytical algorithm of Lane-Emden type equation arising in astrophysics using modified homotopy analysis method. *Computer Physics Communications*, 180: 1116–1124, July 2009.
18. B. Muatjetjeja and C.M. Khalique. Exact solutions of the generalized Lane-Emden equations of the first and second kind. *Pramana*, 77: 545–554, 2011.
19. S. Mall and S. Chakraverty. Chebyshev neural network based model for solving Lane–Emden type equations. *Applied Mathematics and Computation*, 247: 100–114, November 2014.

8

Emden–Fowler Equations

Singular second-order nonlinear initial value problems (IVPs) describe several phenomena in mathematical physics and astrophysics. Many problems in astrophysics may be modeled by second-order ordinary differential equations (ODEs) as proposed by Lane [1]. The Emden–Fowler equation has been studied in detail by Emden [2] and Fowler [3,4]. The general form of the Emden–Fowler equation is written as

$$\frac{d^2y}{dx^2}+\frac{r}{x}\frac{dy}{dx}+af(x)g(y)=h(x), \quad r\geq 0$$
$$\text{subject to initial conditions } y(0)=\alpha, y'(0)=0 \tag{8.1}$$

where
$f(x)$, $h(x)$, and $g(y)$ are functions of x and y, respectively
r, a, and α are constants

For $f(x)=1$, $g(y)=y^n$, and $r=2$, Equation 8.1 reduces to the standard Lane–Emden equations. The Emden–Fowler-type equations are applicable to the theory of stellar structure, thermal behavior of a spherical cloud of gas, isothermal gas spheres, and theory of thermionic currents [5–7]. The solution of differential equations with singularity behavior in various linear and nonlinear IVPs of astrophysics is a challenge. In particular, the present problem of Emden–Fowler equations that has singularity at $x=0$ is also important in practical applications. These equations are difficult to solve analytically, so various techniques based on series solutions such as Adomian decomposition, differential transformation, and perturbation methods, namely, homotopy perturbation, have been employed to solve Emden–Fowler equations. Wazwaz [8–10] used the Adomian decomposition method (ADM) and modified decomposition method for solving Lane–Emden and Emden–Fowler-type equations. Chowdhury and Hashim [11,12] employed the homotopy perturbation method (HPM) to solve singular IVPs of time-independent equations. Ramos [13] solved singular IVPs of ODEs using linearization techniques. Liao [14] used ADM for solving Lane–Emden-type equations. An approximate solution of a differential equation arising in astrophysics using the variational iteration method has been developed by Dehghan and Shakeri [15]. The Emden–Fowler equation has also been solved by utilizing the techniques of Lie and Painleve analysis in Govinder and Leach [16]. An efficient analytical algorithm based on modified homotopy analysis method has been implemented by Singh et al. [17]. Muatjetjeja and Khalique [18] provided an exact solution of the generalized Lane–Emden equations of the first and second kind. Mellin et al. [19] numerically solved general Emden–Fowler equations with two symmetries. Vanani and Aminataei [20] implemented the Pade series solution of Lane–Emden equations. Demir and Sungu [21] approached numerical solutions of nonlinear singular IVPs of Emden–Fowler type using the differential transformation method (DTM) and Maple 11. Kusano and Manojlovic [22] presented the asymptotic behavior of positive solutions of second-order nonlinear ODEs of Emden–Fowler type. Bhrawy and Alofi [23] proposed a shifted Jacobi–Gauss collocation spectral method for solving nonlinear Lane–Emden-type equations. The homotopy analysis method for singular IVPs of Emden–Fowler type has been proposed by Bataineh et al. [24]. In another approach, Muatjetjeja and Khalique [25] developed conservation laws for a generalized coupled bidimensional Lane–Emden system.

Multilayer ANN and single-layer functional link artificial neural network (FLANN) models have been used here to handle homogeneous and nonhomogeneous Emden–Fowler equations. The feed-forward neural network model with error back-propagation algorithm is used for modifying the network parameters and minimizing the computed error function. Initial weights of the ANN model are considered as random.

8.1 Multilayer ANN-Based Solution of Emden–Fowler Equations

The formulation, error computation, and learning algorithm of the multilayer ANN are already described in Section 3.2.2.2 (Equations 3.16 through 3.22). Here, we have considered Emden–Fowler equations with two examples.

Example 8.1

In this example, we take a nonlinear, homogeneous Emden–Fowler equation

$$y'' + \frac{6}{x}y' + 14y = -4y\ln y \quad x \ge 0$$

subject to $y(0)=1,\ y'(0)=0$.

The analytical solution is [13]

$$y(x) = e^{-x^2}$$

We can write the ANN trial solution as

$$y_t(x, p) = 1 + x^2 N(x, p)$$

The network is trained for 10 equidistant points in the given domain. Analytical and traditional (multilayer perceptron [MLP]) ANN solutions are presented in Table 8.1. Figure 8.1 shows a semilogarithmic plot of the error function (between analytical and ANN solutions). The converged ANN is used then to have the results for some testing points. As such, Table 8.2 includes the corresponding results directly by using converged weights.

TABLE 8.1

Comparison between Analytical and ANN Solutions (Example 8.1)

Input Data	Analytical [13]	Traditional ANN
0	1.00000000	1.00000000
0.1	0.99004983	0.99014274
0.2	0.96078943	0.96021042
0.3	0.91393118	0.91302963
0.4	0.85214378	0.85376495
0.5	0.77880078	0.77644671
0.6	0.69767632	0.69755681
0.7	0.61262639	0.61264315
0.8	0.52729242	0.52752822
0.9	0.44485806	0.44502071
1.0	0.36787944	0.36767724

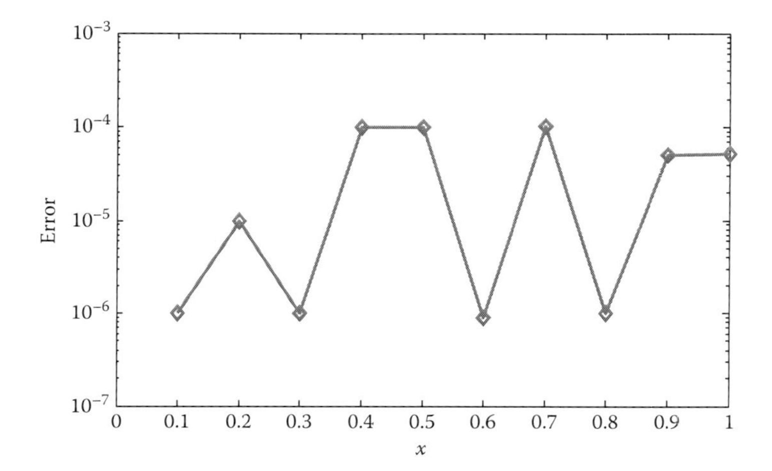

FIGURE 8.1
Semilogarithmic plot of error between analytical and ANN solutions (Example 8.1).

TABLE 8.2

Analytical and ANN Solutions for Testing Points (Example 8.1)

Testing Points	0.173	0.281	0.467	0.650	0.872
Analytical	0.97051443	0.92407596	0.80405387	0.65540625	0.46748687
ANN	0.97052804	0.92419415	0.80387618	0.65579724	0.46739614

Example 8.2

A nonlinear singular IVP of Emden–Fowler type may be written as

$$y'' + \frac{8}{x}y' + xy^2 = x^4 + x^5 \quad x \ge 0$$

with initial conditions $y(0) = 1$, $y'(0) = 0$.

As mentioned in Example 8.1, we have the ANN trial solution

$$y_t(x, p) = 1 + x^2 N(x, p)$$

We train the network for 10 equidistant points in the domain [0, 1] and 7 nodes in the hidden layer. Table 8.3 shows a comparison among numerical solutions obtained by Maple 11, DTM for n = 10 [21], and traditional (MLP) ANN. The comparison between numerical solutions by Maple 11 and ANN is depicted in Figure 8.2. Figure 8.3 shows a semilogarithmic plot of the error (between Maple 11 and ANN).

TABLE 8.3

Comparison among Numerical Solutions Using Maple 11, DTM, and Traditional ANN (Example 8.2)

Input Data	Maple 11 [21]	DTM [21]	Traditional ANN
0	1.0000000	1.00000000	1.00000000
0.1	0.99996668	0.99996668	0.99989792
0.2	0.99973433	0.99973433	1.00020585
0.3	0.99911219	0.99911219	0.99976618
0.4	0.99793933	0.99793933	0.99773922
0.5	0.99612622	0.99612622	0.99652763
0.6	0.99372097	0.99372096	0.99427655
0.7	0.99100463	0.99100452	0.99205860
0.8	0.98861928	0.98861874	0.98867279
0.9	0.98773192	0.98772971	0.98753290
1.0	0.99023588	0.99022826	0.99068174

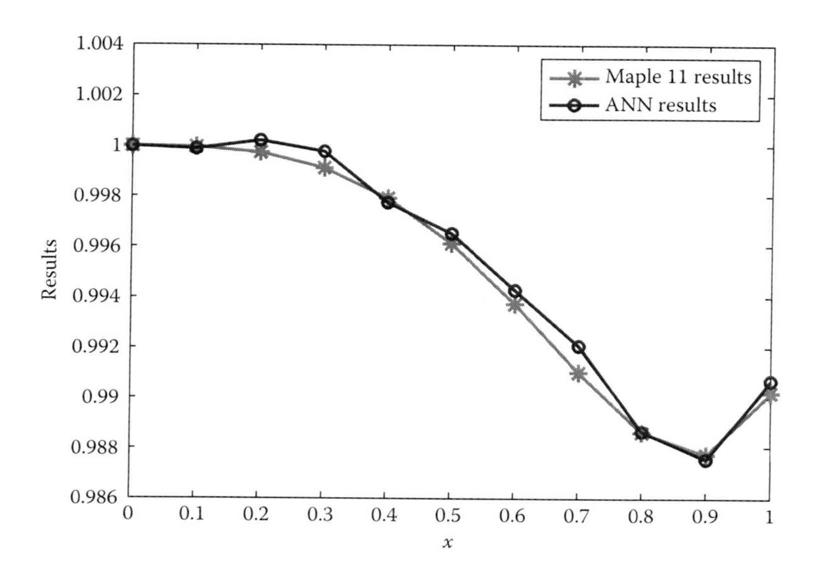

FIGURE 8.2
Plot of numerical solutions using Maple 11 and ANN (Example 8.2).

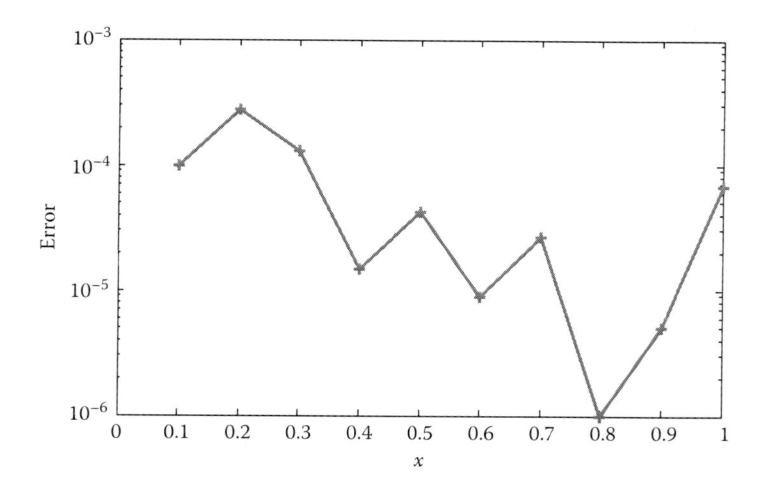

FIGURE 8.3
Semilogarithmic plot of error between Maple 11 and ANN solutions (Example 8.2).

8.2 FLANN-Based Solution of Emden–Fowler Equations

In this section, the single-layer FLANN model has been used to solve singular nonlinear ODEs of Emden–Fowler-type equations [26,27].

The Chebyshev neural network (ChNN) model has been used for solving Emden–Fowler equations. In this regard, the structure, formulation, and error computation of ODEs using the ChNN model have already been discussed in Section 5.1.1.2. The ChNN trial solution for a second-order IVP is same as the trial solution of the multilayer ANN. The learning algorithm for ChNN is discussed in Section 5.1.1.3. Again, two examples have been considered in the subsequent paragraphs.

Example 8.3

Now let us consider a nonhomogeneous Emden–Fowler equation

$$y'' + \frac{8}{x}y' + xy = x^5 - x^4 + 44x^2 - 30x \quad x \geq 0$$

with initial conditions $y(0) = 0$, $y'(0) = 0$.

The related ChNN trial solution is

$$y_t(x, p) = x^2 N(x, p)$$

Ten equidistant points in [0, 1] and six weights with respect to the first six Chebyshev polynomials are considered. A comparison of analytical and ChNN solutions has been presented in Table 8.4. This comparison is also

TABLE 8.4

Comparison between Analytical and ChNN Solutions (Example 8.3)

Input Data	Analytical [12]	ChNN [26]
0	0	0
0.1	−0.00090000	−0.00058976
0.2	−0.00640000	−0.00699845
0.3	−0.01890000	−0.01856358
0.4	−0.03840000	−0.03838897
0.5	−0.06250000	−0.06318680
0.6	−0.08640000	−0.08637497
0.7	−0.10290000	−0.10321710
0.8	−0.10240000	−0.10219490
0.9	−0.07290000	−0.07302518
1.0	0.00000000	0.00001103

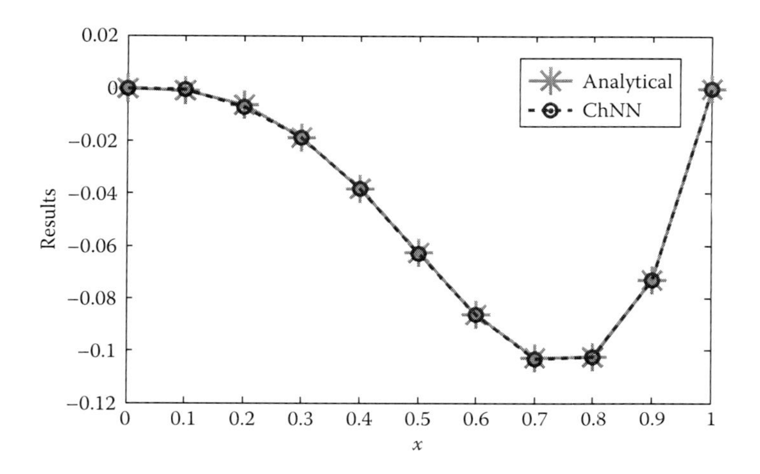

FIGURE 8.4
Plot of analytical and ChNN solutions (Example 8.3).

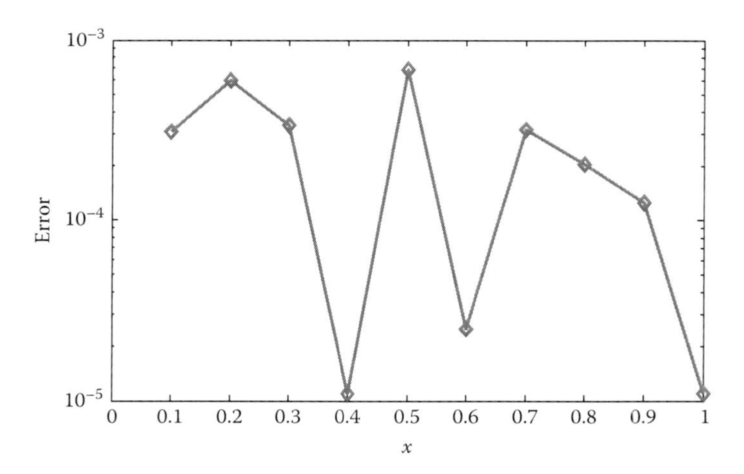

FIGURE 8.5
Semilogarithmic plot of error between analytical and ChNN solutions (Example 8.3).

depicted in Figure 8.4. A semilogarithmic plot of the error function between analytical and ChNN solutions is presented in Figure 8.5. Finally, results for some testing points are shown in Table 8.5. This testing is done to check whether the converged ChNN can give results directly by inputting the points that were not taken during training.

TABLE 8.5

Analytical and ChNN Solutions for Testing Points (Example 8.3)

Testing Points	0.154	0.328	0.561	0.732	0.940
Analytical	−0.00308981	−0.02371323	−0.07750917	−0.10511580	−0.04983504
ChNN	−0.00299387	−0.02348556	−0.07760552	−0.10620839	−0.04883402

Example 8.4

Finally, we consider a nonlinear Emden–Fowler equation

$$y'' + \frac{3}{x}y' + 2x^2y^2 = 0$$

with initial conditions $y(0)=1$, $y'(0)=0$.

The ChNN trial solution in this case is represented as

$$y_t(x, p) = 1 + x^2N(x, p)$$

Again, the network is trained for 10 equidistant points. Table 8.6 includes the comparison among solutions obtained by Maple 11, DTM for n = 10 [21], and ChNN. Figure 8.6 shows the comparison between numerical solutions by Maple 11 and ChNN. Finally, a semilogarithmic plot of the error (between Maple 11 and ChNN solutions) is presented in Figure 8.7.

TABLE 8.6

Comparison among Numerical Solutions by Maple 11, DTM for n = 10, and ChNN (Example 8.4)

Input Data	Maple 11 [21]	DTM [21]	ChNN [26]
0	1.00000000	1.00000000	1.00000000
0.1	0.99999166	0.99999166	0.99989166
0.2	0.99986667	0.99986667	0.99896442
0.3	0.99932527	0.99932527	0.99982523
0.4	0.99786939	0.99786939	0.99785569
0.5	0.99480789	0.99480794	0.99422605
0.6	0.98926958	0.98926998	0.98931189
0.7	0.98022937	0.98023186	0.98078051
0.8	0.96655340	0.96656571	0.96611140
0.9	0.94706857	0.94711861	0.94708231
1.0	0.92065853	0.92083333	0.92071830

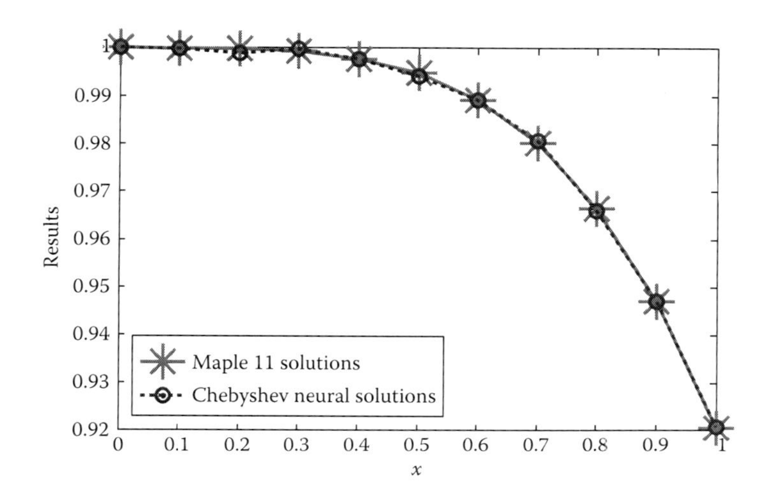

FIGURE 8.6
Plot of Maple 11 and ChNN solutions (Example 8.4).

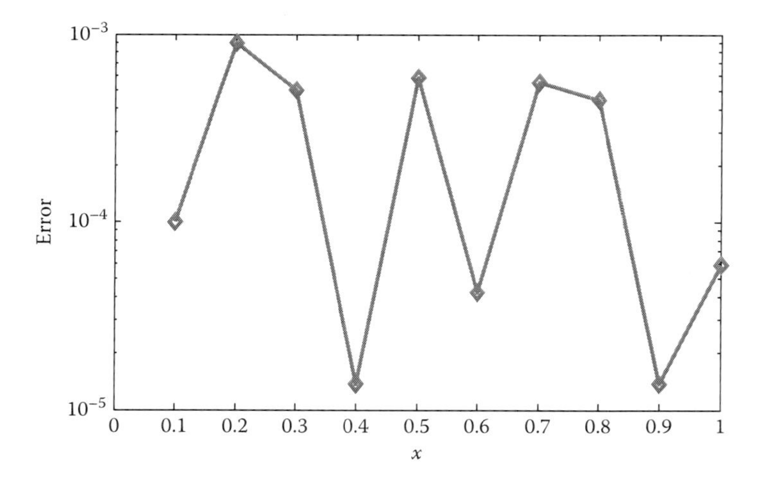

FIGURE 8.7
Semilogarithmic plot of error between Maple 11 and ChNN solutions (Example 8.4).

References

1. J.H. Lane. On the theoretical temperature of the sun under the hypothesis of a gaseous mass maintaining its volume by its internal heat and depending on the laws of gases known to terrestrial experiment. *The American Journal of Science and Arts*, 2nd series, 4: 57–74, 1870.
2. R. Emden. *Gaskugeln Anwendungen der mechanischen Warmen-theorie auf Kosmologie and meteorologische Probleme*. Teubner, Leipzig, Germany, 1907.

3. R.H. Fowler. The form near infinity of real, continuous solutions of a certain differential equation of the second order. *Quarterly Journal of Mathematics (Oxford)*, 45: 341–371, 1914.
4. R.H. Fowler. Further studies of Emden's and similar differential equations. *Quarterly Journal of Mathematics (Oxford)*, 2: 259–288, 1931.
5. H.T. Davis. *Introduction to Nonlinear Differential and Integral Equations*. Dover Publications Inc., New York, 1962.
6. S. Chandrasekhar. *Introduction to Study of Stellar Structure*. Dover Publications Inc., New York, 1967.
7. B.K. Datta. Analytic solution to the Lane-Emden equation. *Nuovo Cimento*, 111B: 1385–1388, 1996.
8. A.M. Wazwaz. A new algorithm for solving differential equation Lane–Emden type. *Applied Mathematics and Computation*, 118: 287–310, 2001.
9. A.M. Wazwaz. Adomian decomposition method for a reliable treatment of the Emden–Fowler equation. *Applied Mathematics and Computation*, 161: 543–560, 2005.
10. A.M. Wazwaz. The modified decomposition method for analytical treatment of differential equations. *Applied Mathematics and Computation*, 173: 165–176, 2006.
11. M.S.H. Chowdhury and I. Hashim. Solutions of a class of singular second order initial value problems by homotopy-perturbation method. *Physics Letters A*, 365: 439–447, 2007.
12. M.S.H. Chowdhury and I. Hashim. Solutions of Emden-Fowler equations by homotopy-perturbation method. *Nonlinear Analysis: Real Word Applications*, 10: 104–115, 2009.
13. J.I. Ramos. Linearization techniques for singular initial-value problems of ordinary differential equations. *Applied Mathematics and Computation*, 161: 525–542, 2005.
14. S.J. Liao. A new analytic algorithm of Lane–Emden type equations. *Applied Mathematics and Computation*, 142: 1–16, 2003.
15. M. Dehghan and F. Shakeri. Approximate solution of a differential equation arising in astrophysics using the variational iteration method. *New Astronomy*, 13: 53–59, 2008.
16. K.S. Govinder and P.G.L. Leach. Integrability analysis of the Emden-Fowler equation. *Journal of Nonlinear Mathematical Physics*, 14: 435–453, 2007.
17. O.P. Singh, R.K. Pandey, and V.K. Singh. Analytical algorithm of Lane-Emden type equation arising in astrophysics using modified homotopy analysis method. *Computer Physics Communications*, 180: 1116–1124, 2009.
18. B. Muatjetjeja and C.M. Khalique. Exact solutions of the generalized Lane-Emden equations of the first and second kind. *Pramana*, 77: 545–554, 2011.
19. C.M. Mellin, F.M. Mahomed, and P.G.L. Leach. Solution of generalized Emden-Fowler equations with two symmetries. *International Journal of NonLinear Mechanics*, 29: 529–538, 1994.
20. S.K. Vanani and A. Aminataei. On the numerical solution of differential equations of Lane-Emden type. *Computers and Mathematics with Applications*, 59: 2815–2820, 2010.
21. H. Demir and I.C. Sungu. Numerical solution of a class of nonlinear Emden-Fowler equations by using differential transformation method. *Journal of Arts and Science*, 12: 75–81, 2009.

22. T. Kusano and J. Manojlovic. Asymptotic behavior of positive solutions of sub linear differential equations of Emden–Fowler type. *Computers and Mathematics with Applications*, 62: 551–565, 2011.
23. A.H. Bhrawy and A.S. Alofi. A Jacobi-Gauss collocation method for solving nonlinear Lane-Emden type equations. *Communications in Nonlinear Science and Numerical Simulation*, 17: 62–70, 2012.
24. A.S. Bataineh, M.S.M. Noorani, and I. Hashim. Homotopy analysis method for singular initial value problems of Emden-Fowler type. *Communications in Nonlinear Science and Numerical Simulation*, 14: 1121–1131, 2009.
25. B. Muatjetjeja and C.M. Khalique. Conservation laws for a generalized coupled bi dimensional Lane–Emden system. *Communications in Nonlinear Science and Numerical Simulation*, 18: 851–857, 2013.
26. S. Mall and S. Chakraverty. Numerical solution of nonlinear singular initial value problems of Emden–Fowler type using Chebyshev neural network method. *Neurocomputing*, 149: 975–982, 2015.
27. S. Mall and S. Chakraverty. Multi layer versus functional link single layer neural network for solving nonlinear singular initial value problems. *Third International Symposium on Women Computing and Informatics (WCI-2015)*, SCMS College, Kochi, Kerala, India, August 10–13, Published in Association for Computing Machinery (ACM) Proceedings, pp. 678–683, 2015.

9

Duffing Oscillator Equations

Duffing oscillators play a crucial role in applied mathematics, physics, and engineering problems. The nonlinear Duffing oscillator equations have various engineering applications, namely, nonlinear vibration of beams and plates [1], magneto-elastic mechanical systems [2], fluid flow–induced vibration [3], etc. The solution of these problems has been a recent research topic because most of them do not have analytical solutions.

Thus, various numerical techniques and perturbation methods have been used to handle Duffing oscillator equations. Nourazar and Mirzabeigy [4] employed the modified differential transformation method (MDTM) to solve Duffing oscillator with damping effect. An approximate solution of force-free Duffing–van der Pol oscillator equations using the homotopy perturbation method (HPM) has been developed by Khan et al. [5]. Younesian [6] provided free vibration analysis of strongly nonlinear generalized Duffing oscillators using He's variational approach and HPM. Duffing–van der Pol oscillator equation has been solved by Chen and Liu [7] using Liao's homotopy analysis method. Kimiaeifar et al. [8] proposed homotopy analysis method for solving van der Pol–Duffing oscillators. In [9], Akbarzade and Ganji have implemented homotopy perturbation and variational methods for the solution of nonlinear cubic-quintic Duffing oscillators. Mukherjee et al. [10] evaluated the solution of the Duffing–van der Pol equation by differential transformation method (DTM). Njah and Vincent [11] presented chaos synchronization between single- and double-well Duffing–van der Pol oscillators using the active control technique. Ganji et al. [12] used He's energy balance method to solve strongly nonlinear Duffing oscillators equations. Akbari et al. [13] solved van der Pol, Rayleigh, and Duffing equations using the algebraic method. In [14], Hu and Wen applied the Duffing oscillator for extracting the features of early mechanical failure signal.

9.1 Governing Equation

The general form of the damped Duffing oscillator equation is expressed as

$$\frac{d^2x}{dt^2} + \alpha\frac{dx}{dt} + \beta x + \gamma x^3 = F\cos\omega t \quad \alpha \ge 0 \tag{9.1}$$

with initial conditions $x(0) = a, x'(0) = b$

where

- α represents the damping coefficient
- F and ω denote the magnitude of periodic force and the frequency of this force, respectively
- t is the periodic time

Equation 9.1 reduces to the unforced damped Duffing oscillator equation when $F = 0$.

Our aim is to discuss the solution of unforced and forced Duffing oscillator equations using the single-layer functional link artificial neural network (FLANN) method. In this chapter, we have considered single-layer simple orthogonal polynomial–based neural network (SOPNN) and Hermite neural network (HeNN) models to handle these equations.

It may be noted that structure, formulation, error computation, and learning algorithm of the SOPNN model are already described in Sections 5.1.4.1 through 5.1.4.3, respectively. Similarly, the architecture of HeNN, and its formulation and learning algorithm are also included in Sections 5.1.3.1 and 5.1.3.2, respectively.

9.2 Unforced Duffing Oscillator Equations

This section includes unforced damped Duffing oscillator equations to show the powerfulness of the SOPNN method.

Example 9.1

Let us take a force-free damped Duffing oscillator problem [4] with $\alpha = 0.5$, $\beta = \gamma = 25$, $a = 0.1$, and $b = 0$.

Accordingly, we have

$$\frac{d^2x}{dt^2} + 0.5\frac{dx}{dt} + 25x + 25x^3 = 0$$

subject to initial conditions $x(0) = 0.1$, $x'(0) = 0$.

As discussed in Section 5.1.4.2, we may write the SOPNN trial solution as

$$x_\phi(t, p) = 0.1 + t^2 N(t, p)$$

The network has been trained for 50 points in the domain [0, 5] and 6 weights with respect to the first 6 simple orthogonal polynomials. Table 9.1 shows a comparison among numerical results obtained by the MDTM [4] by the Pade

TABLE 9.1

Comparison among MDTM and SOPNN Results (Example 9.1)

Input Data t (Time)	MDTM by the Pade Approximate of [3/3] [4]	Real Part of MDTM by the Pade Approximate of [4/4] [4]	SOPNN
0	0.0998	0.1000	0.1000
0.1000	0.0876	0.0853	0.0842
0.2000	0.0550	0.0511	0.0512
0.3000	0.0110	0.0064	0.0063
0.4000	−0.0331	−0.0380	−0.0377
0.5000	−0.0665	−0.0712	−0.0691
0.6000	−0.0814	−0.0854	−0.0856
0.7000	−0.0751	−0.0782	−0.0764
0.8000	−0.0502	−0.0528	−0.0530
0.9000	−0.0136	−0.0161	−0.0173
1.0000	0.0249	0.0229	0.0224
1.1000	0.0560	0.0547	0.0539
1.2000	0.0722	0.0716	0.0713
1.3000	0.0704	0.0703	0.0704
1.4000	0.0518	0.0523	0.0519
1.5000	0.0219	0.0227	0.0215
1.6000	−0.0115	−0.0110	−0.0109
1.7000	−0.0399	−0.0406	−0.0394
1.8000	−0.0567	−0.0589	−0.0574
1.9000	−0.0582	−0.0620	−0.0624
2.0000	−0.0449	−0.0501	−0.0489
2.1000	−0.0207	−0.0268	−0.0262
2.2000	0.0078	0.0018	−0.0018
2.3000	0.0336	0.0287	0.0279
2.4000	0.0503	0.0473	0.0477
2.5000	0.0543	0.0536	0.0534
2.6000	0.0452	0.0467	0.0461
2.7000	0.0260	0.0290	0.0289
2.8000	0.0017	0.0051	0.0052
2.9000	−0.0212	−0.0189	−0.0191
3.0000	−0.0374	−0.0372	−0.0375
3.1000	−0.0431	−0.0456	−0.0487
3.2000	−0.0375	−0.0426	−0.0404
3.3000	−0.0224	−0.0295	−0.0310
3.4000	−0.0021	−0.0101	−0.0135
3.5000	0.0182	0.0109	0.0147
3.6000	0.0335	0.0283	0.0308
3.7000	0.0404	0.0380	0.0400
3.8000	0.0374	0.0380	0.0388

(Continued)

TABLE 9.1 (*Continued*)

Comparison among MDTM and SOPNN Results (Example 9.1)

Input Data t (Time)	MDTM by the Pade Approximate of [3/3] [4]	Real Part of MDTM by the Pade Approximate of [4/4] [4]	SOPNN
3.9000	0.0259	0.0290	0.0279
4.0000	0.0090	0.0134	0.0113
4.1000	−0.0088	−0.0047	−0.0076
4.2000	−0.0230	−0.0208	−0.0221
4.3000	−0.0305	−0.0311	−0.0309
4.4000	−0.0296	−0.0334	−0.0325
4.5000	−0.0210	−0.0275	−0.0258
4.6000	−0.0072	−0.0154	−0.0124
4.7000	0.0082	−0.0001	0.0032
4.8000	0.0213	0.0145	0.0193
4.9000	0.0290	0.0248	0.0250
5.0000	0.0297	0.0287	0.0288

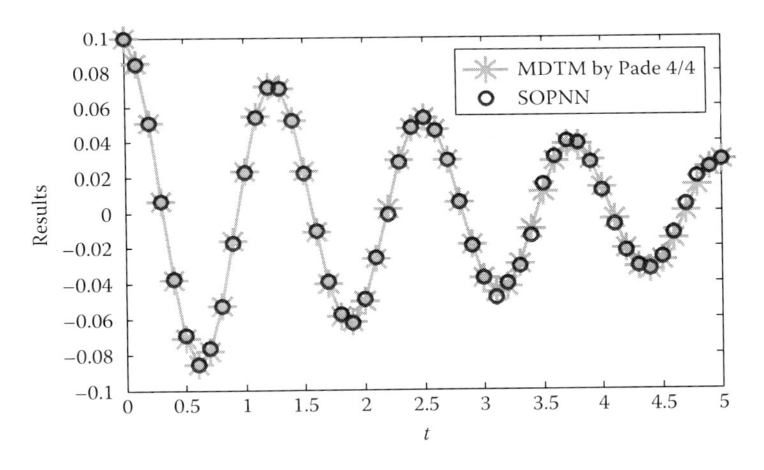

FIGURE 9.1
Plot of MDTM and SOPNN results (Example 9.1).

approximate of [3/3], real part of MDTM by the Pade approximate of [4/4], and SOPNN. The comparison between results by real part of MDTM [4] and SOPNN is depicted in Figure 9.1. The plot of the error function (between MDTM and SOPNN) has also been shown in Figure 9.2. The phase plane diagram by SOPNN is presented in Figure 9.3.

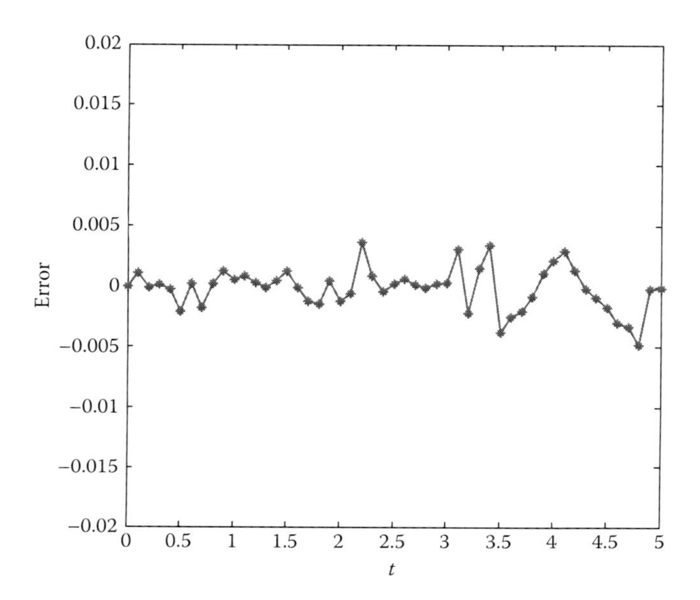

FIGURE 9.2
Error plot between MDTM and SOPNN results (Example 9.1).

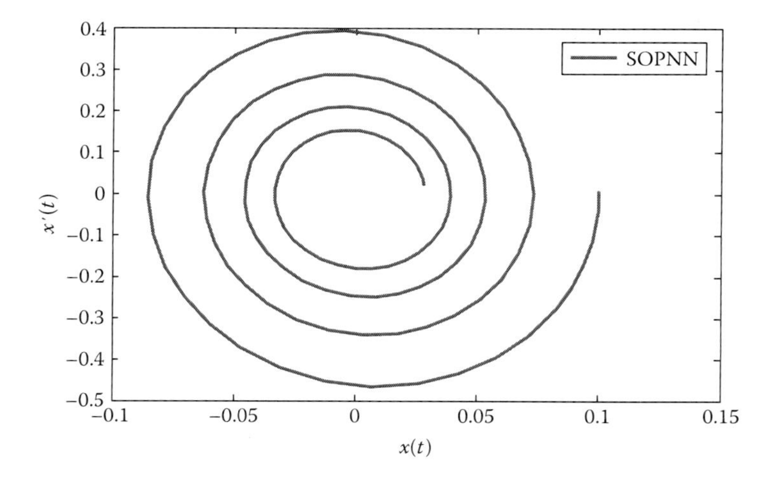

FIGURE 9.3
Phase plane plot by SOPNN (Example 9.1).

TABLE 9.2

Comparison between MDTM and SOPNN Results (Example 9.2)

Input Data *t* (Time)	MDTM [4]	SOPNN
0	−0.2000	−0.2000
0.1000	0.0031	0.0034
0.2000	0.1863	0.1890
0.3000	0.3170	0.3172
0.4000	0.3753	0.3745
0.5000	0.3565	0.3587
0.6000	0.2714	0.2711
0.7000	0.1427	0.1400
0.8000	−0.0010	−0.0014
0.9000	−0.1307	−0.1299
1.0000	−0.2234	−0.2250
1.1000	−0.2648	−0.2639
1.2000	−0.2519	−0.2510
1.3000	−0.1923	−0.1930
1.4000	−0.1016	−0.1013
1.5000	−0.0002	−0.0005
1.6000	0.0916	0.0919
1.7000	0.1574	0.1570
1.8000	0.1869	0.1872
1.9000	0.1781	0.1800
2.0000	0.1362	0.1360
2.1000	0.0724	0.0726
2.2000	0.0007	0.0006
2.3000	−0.0642	−0.0661
2.4000	−0.1108	−0.1095
2.5000	−0.1319	−0.1309
2.6000	−0.1259	−0.1262
2.7000	−0.0965	−0.0969
2.8000	−0.0515	−0.0519
2.9000	−0.0009	−0.0003
3.0000	0.0450	0.0456
3.1000	0.0781	0.0791
3.2000	0.0931	0.0978
3.3000	0.0890	0.0894
3.4000	0.0684	0.0671
3.5000	0.0367	0.0382
3.6000	0.0010	0.0021
3.7000	−0.0315	−0.0338
3.8000	−0.0550	−0.0579

(*Continued*)

TABLE 9.2 (*Continued*)

Comparison between MDTM and SOPNN Results (Example 9.2)

Input Data *t* (Time)	MDTM [4]	SOPNN
3.9000	−0.0657	−0.0666
4.0000	−0.0629	−0.0610
4.1000	−0.0484	−0.0501
4.2000	−0.0261	−0.0285
4.3000	−0.0009	−0.0005
4.4000	0.0221	0.0210
4.5000	0.0387	0.0400
4.6000	0.0464	0.0472
4.7000	0.0445	0.0439
4.8000	0.0343	0.0358
4.9000	0.0186	0.0190
5.0000	0.0008	0.0006

Example 9.2

Next, we have taken the unforced Duffing oscillator problem with $\alpha = 1$, $\beta = 20$, $\gamma = 2$, $a = -0.2$, and $b = 2$.

The differential equation may be written as

$$\frac{d^2x}{dt^2} + \frac{dx}{dt} + 20x + 2x^3 = 0$$

subject to $x(0) = -0.2$, $x'(0) = 2$.

The SOPNN trial solution, in this case, is represented as

$$x_\phi(t, p) = -0.2 + 2t + t^2 N(t, p)$$

Again, the network is trained with 50 equidistant points in the interval [0, 5] and the first 6 simple orthogonal polynomials. Table 9.2 includes a comparison between results of the real part of MDTM [4] by the Pade approximate of [4/4] and SOPNN. Figure 9.4 shows the comparison of results between [4] and SOPNN. The plot of the error is depicted in Figure 9.5. Figure 9.6 depicts the phase plane diagram.

9.3 Forced Duffing Oscillator Equations

Here, we have considered the Duffing oscillator equation with periodic force. Two example problems have been solved using the single-layer HeNN method.

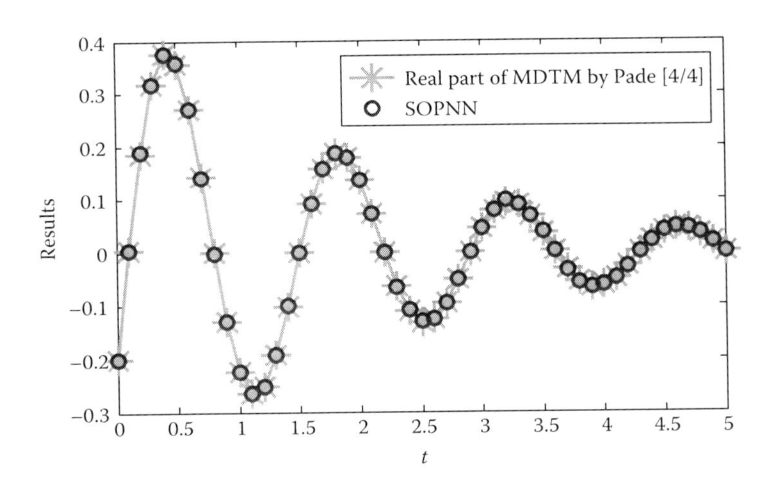

FIGURE 9.4
Plot of MDTM and SOPNN results (Example 9.2).

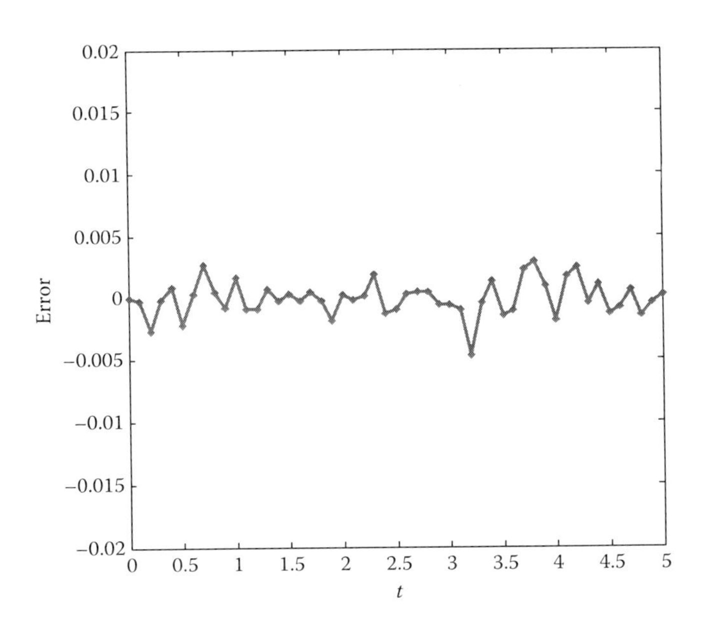

FIGURE 9.5
Error plot between MDTM and SOPNN results (Example 9.2).

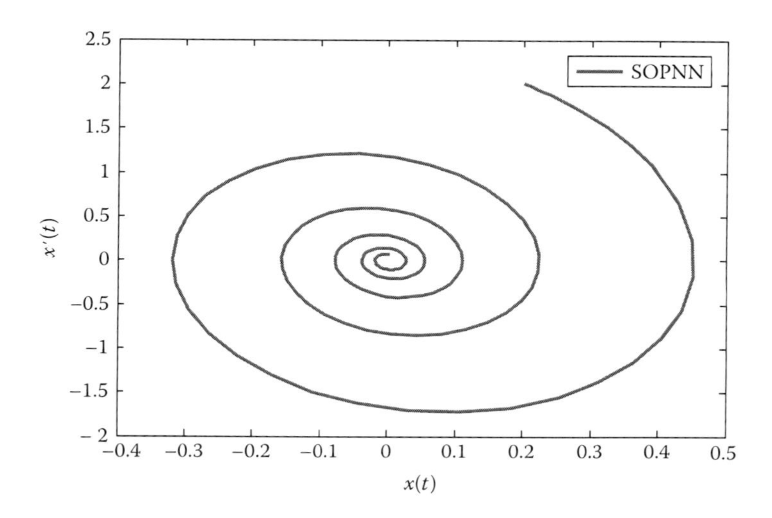

FIGURE 9.6
Phase plane plot by SOPNN (Example 9.2).

Example 9.3

A Duffing oscillator equation with periodic force 0.2 [13]

$$\frac{d^2x}{dt^2} + 0.3^2(x + 0.2^2x^3) = 0.2\sin 2t$$

with initial conditions $x(0) = 0.15$, $x'(0) = 0$.

As discussed in Section 5.1.3.1, we can write the HeNN trial solution as

$$x_{\mathrm{He}}(t, p) = 0.15 + t^2 N(t, p)$$

Here, the network is trained for 225 equidistant points in the time interval [0, 45]. Seven weights from the input layer to the output layer with respect to the first seven Hermite polynomials have been considered for training. Figure 9.7 shows a comparison of numerical results $x(t)$ among the Runge–Kutta method (RKM), HeNN, and the algebraic method (AGM) [13]. Figures 9.8 and 9.9 depict the phase plane diagram between displacement and velocity using HeNN and RKM, respectively. From Figure 9.7, one can observe that the results obtained by RKM and algebraic method (AGM) agree exactly at all points with HeNN results.

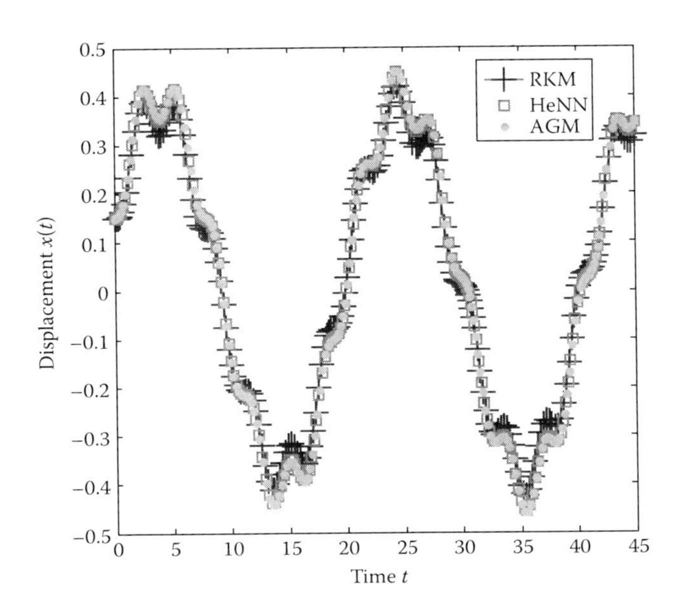

FIGURE 9.7
Comparison among RKM, HeNN, and AGM [13] results (Example 9.3).

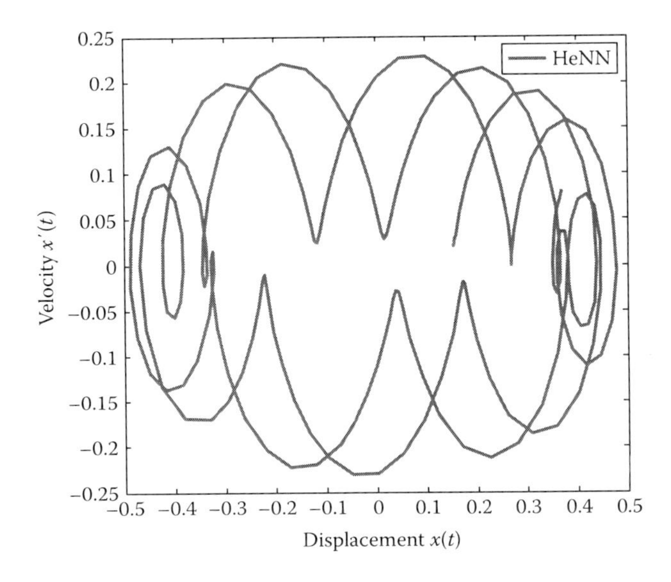

FIGURE 9.8
Phase plane plot by HeNN (Example 9.3).

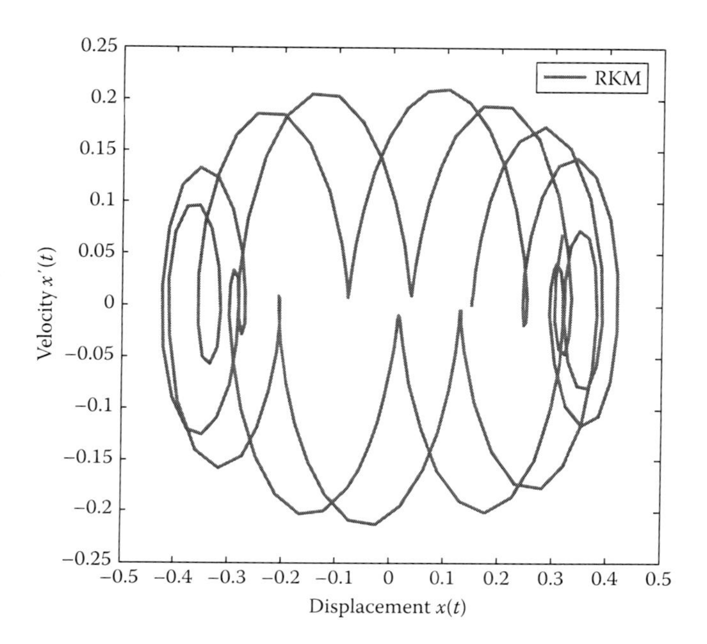

FIGURE 9.9
Phase plane plot by RKM (Example 9.3).

Example 9.4

Here, we have considered a Duffing oscillator equation used for extracting the features of early mechanical failure signal and detect early fault [14]

$$\frac{dx^2}{d^2t} + \delta\frac{dx}{dt} - x + x^3 = \gamma\cos t + s(t)$$
$$\text{subject to } x(0) = 1.0, x'(0) = 1.0$$

where $\delta = 0.5$, $\gamma = 0.8275$ (amplitude of external exciting periodic force), and $s(t) = 0.0005\cos t$ (frequency of external weak signal). We have considered the signal with very low frequency ($\omega = 1$) for the present problem.

The first term on the right-hand side of this equation is the reference signal and the second term is the signal to be detected.

For this problem, we may write the HeNN trial solution as

$$x_{\text{He}}(t, p) = 1.0 + 1.0t + t^2N(t, p)$$

In this application problem, we have considered time t from 0 to 500 s, with a step length $h = 0.5$ and seven weights with respect to the first seven Hermite polynomials. In [14], the authors solved the problem by RKM. The time series plots by RKM [14] and HeNN have been shown in Figures 9.10 and 9.11,

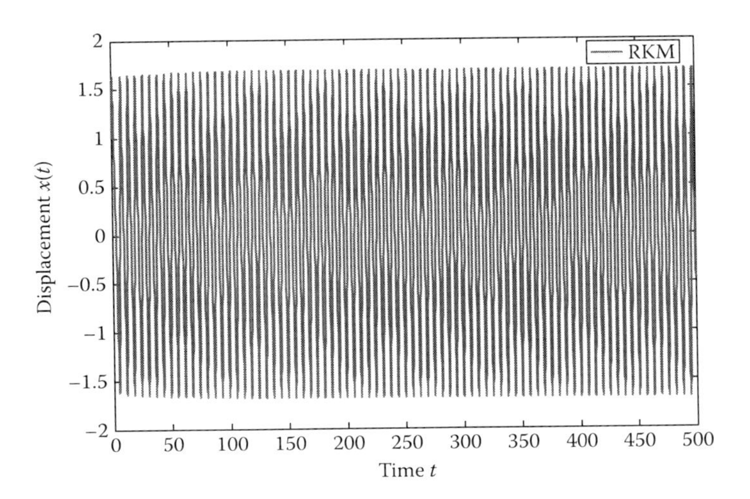

FIGURE 9.10
Time series plot by RKM [14] (Example 9.4).

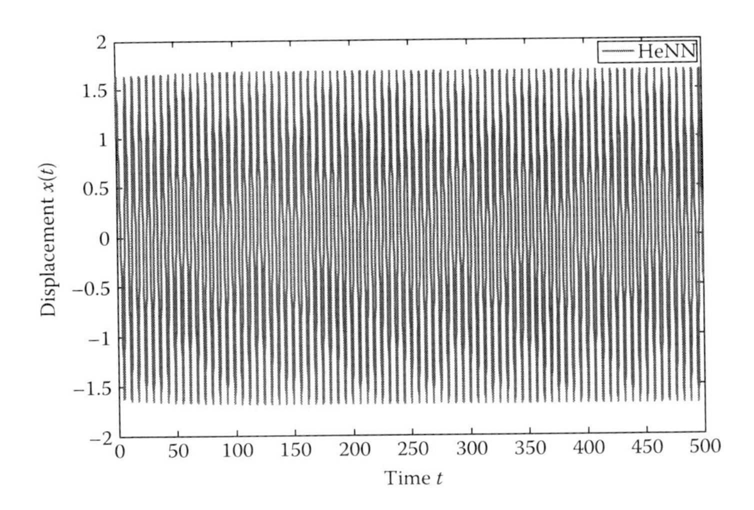

FIGURE 9.11
Time series plot by HeNN (Example 9.4).

respectively. The phase plane plots obtained by RKM [14] and HeNN method for $\gamma=0.8275$ have been depicted in Figures 9.12 and 9.13, respectively. Similarly, phase plane plots for $\gamma=0.828$ by RKM [14] and HeNN are depicted in Figures 9.14 and 9.15, respectively. The phase trajectories, namely, Figures 9.12 and 9.13, are in a chaotic state. One can observe that the width of the chaotic contour becomes narrow with the increase in the value of γ.

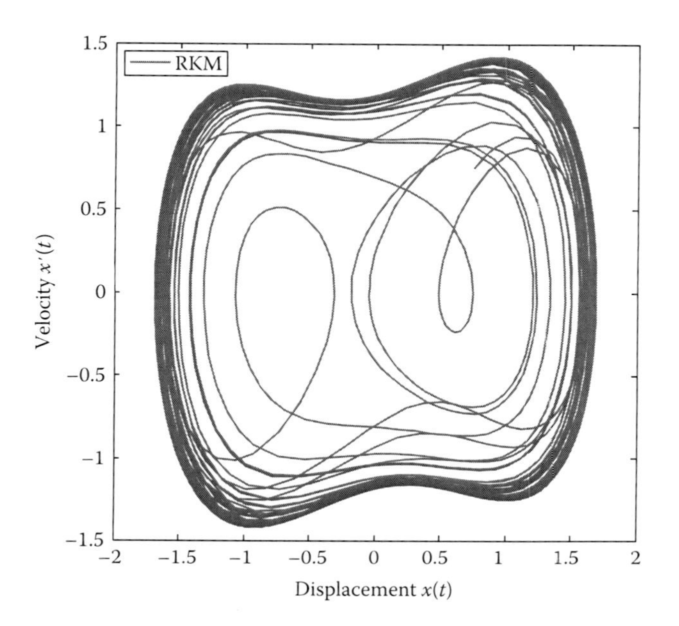

FIGURE 9.12
Phase plane plot by RKM [14] for $\gamma = 0.8275$ (Example 9.4).

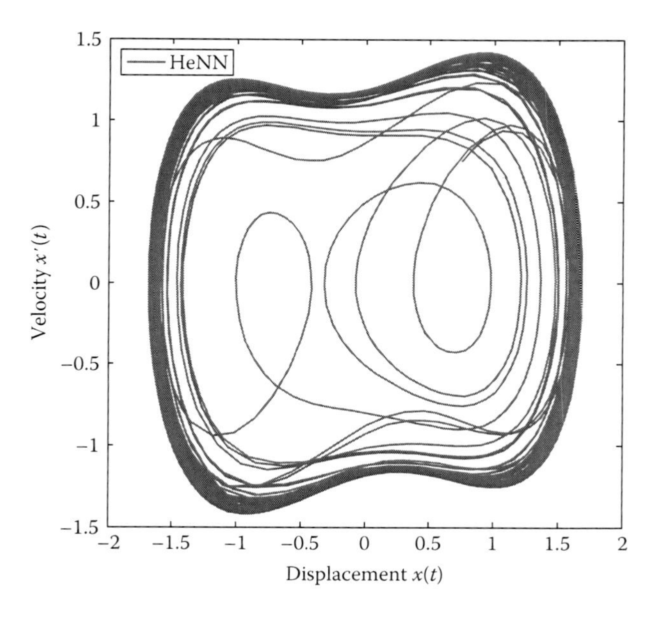

FIGURE 9.13
Phase plane plot by HeNN for $\gamma = 0.8275$ (Example 9.4).

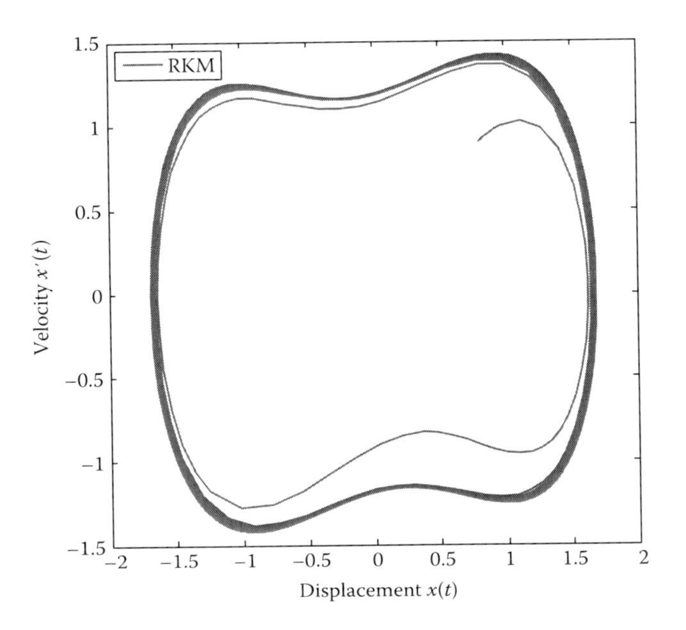

FIGURE 9.14
Phase plane plot by RKM [14] for $\gamma = 0.828$ (Example 9.4).

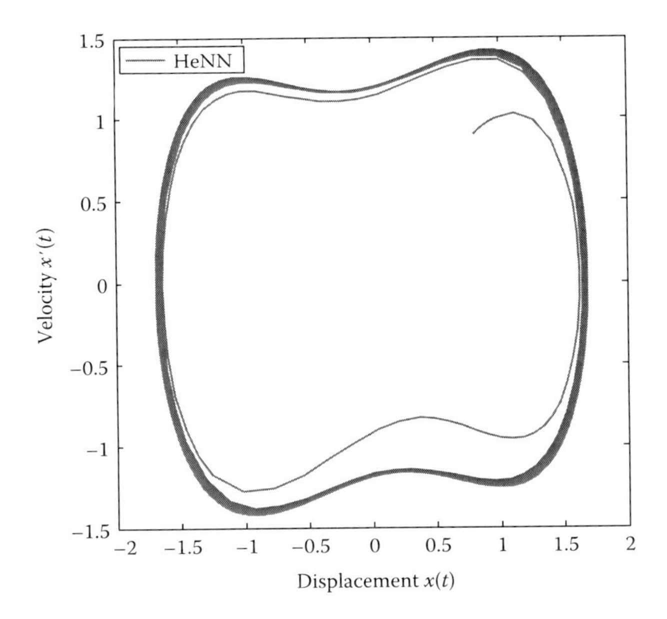

FIGURE 9.15
Phase plane plot by HeNN for $\gamma = 0.828$ (Example 9.4).

In Figures 9.14 and 9.15, the orbit becomes a large periodic motion. From these figures, we can observe that when the amplitude of force γ changes from 0.8275 to 0.828, the chaotic state is gradually replaced by periodic motion [14].

We can compare the solutions of Figures 9.10, 9.12, and 9.14 (obtained by RKM [14]) with Figures 9.11, 9.13, and 9.15 (obtained by the present method), respectively, which are found to be in excellent agreement.

References

1. M.T. Ahmadian, M. Mojahedi, and H. Moeenfard. Free vibration analysis of a nonlinear beam using homotopy and modified Lindstedt–Poincare methods. *Journal of Solid Mechanics*, 1: 29–36, 2009.
2. J. Guckenheimer and P. Holmes. *Nonlinear Oscillations, Dynamical Systems and Bifurcations of Vector Fields*. Springer-Verlag, New York, 1983.
3. N. Srinil and H. Zanganeh. Modelling of coupled cross-flow/in-line vortex-induced vibrations using double Duffing and van der Pol oscillators. *Ocean Engineering*, 53: 83–97, 2012.
4. S. Nourazar and A. Mirzabeigy. Approximate solution for nonlinear Duffing oscillator with damping effect using the modified differential transform method. *Scientia Iranica B*, 20: 364–368, 2013.
5. N.A. Khan, M. Jamil, A.S. Anwar, and N.A. Khan. Solutions of the force-free Duffing-van der pol oscillator equation. *International Journal of Differential Equations*, 2011: 1–9, 2011.
6. D. Younesian, H. Askari, Z. Saadatnia, and M.K. Yazdi. Free vibration analysis of strongly nonlinear generalized Duffing oscillators using He's variational approach and homotopy perturbation method. *Nonlinear Science Letters A*, 2: 11–16, 2011.
7. Y.M. Chen and J.K. Liu. Uniformly valid solution of limit cycle of the Duffing–van der Pol equation. *Mechanics Research Communications*, 36: 845–850, 2009.
8. A. Kimiaeifar, A.R. Saidi, G.H. Bagheri, M. Rahimpour, and D.G. Domairr. Analytical solution for Van der Pol–Duffing oscillators. *Chaos, Solitons and Fractals*, 42: 2660–2666, 2009.
9. M. Akbarzade and D.D. Ganji. Coupled method of homotopy perturbation method and variational approach for solution to Nonlinear Cubic-Quintic Duffing oscillator. *Advances in Theoretical and Applied Mechanics*, 3: 329–337, 2010.
10. S. Mukherjee, B. Roy, and S. Dutta. Solution of the Duffing–van der Pol oscillator equation by a differential transform method. *Physica Scripta*, 83: 1–12, 2010.
11. A.N. Njah and U.E. Vincent. Chaos synchronization between single and double wells Duffing Van der Pol oscillators using active control. *Chaos, Solitons and Fractals*, 37: 1356–1361, 2008.
12. D.D. Ganji, M. Gorji, S. Soleimani, and M. Esmaeilpour. Solution of nonlinear cubic-quintic Duffing oscillators using He's Energy Balance Method. *Journal of Zhejiang University—Science A*, 10(9): 1263–1268, 2009.

13. M.R. Akbari, D.D. Ganji, A. Majidian, and A.R. Ahmadi. Solving nonlinear differential equations of Van der pol. Rayleigh and Duffing by AGM. *Frontiers of Mechanical Engineering*, 9(2): 177–190, 2014.
14. N.Q. Hu and X.S. Wen. The application of Duffing oscillator in characteristic signal detection of early fault. *Journal of Sound and Vibration*, 268: 917–931, 2003.

10

Van der Pol–Duffing Oscillator Equation

The Van der Pol–Duffing oscillator equation is a classical nonlinear oscillator, which is a very useful mathematical model for understanding different engineering problems. This equation is widely used to model various physical problems, namely, electrical circuits, electronics, mechanics, etc. [1]. The Van der Pol oscillator equation was proposed by a Dutch scientist Balthazar Van der Pol [2], which describes triode oscillations in electrical circuits. The Van der Pol–Duffing oscillator is a classical example of a self-oscillatory system and is now considered as a very important model to describe a variety of physical systems. Also this equation describes self-sustaining oscillations in which energy is fed into small oscillations and removed from large oscillations. The nonlinear Duffing oscillator equation and the Van der Pol–Duffing oscillator equations are difficult to solve analytically. In recent years, various types of numerical and perturbation techniques, such as Euler, Runge–Kutta, homotopy perturbation, linearization, and variational iteration methods, have been used to solve the nonlinear equation.

In this regard, the linearization method has been employed by Motsa and Sibanda [3] for solving Duffing and Van der Pol equations. Akbari et al. [4] solved Van der Pol, Rayleigh, and Duffing equations using the algebraic method. In [5], Nourazar and Mirzabeigy evaluated the numerical solution of the Duffing oscillator equation by the modified differential transformation method. An approximate solution of Van der Pol–Duffing equations using the differential transformation method has been studied by Mukherjee et al. [6]. Kimiaeifar et al. [7] proposed the homotopy analysis method for solving single-well, double-well, and double-hump Van der Pol–Duffing oscillator equations. In another work, Njah and Vincent [8] presented chaos synchronization between single- and double-well Duffing–Van der Pol oscillators using the active control technique. An approximate solution of the classical Van der Pol equation using He's parameter expansion method has been developed by Molaei and Kheybari [9]. Sweilam and Al-Bar [10] implemented the parameter expansion method for the coupled Van der Pol oscillator equations. Zhang and Zeng [11] have used a segmenting recursion method to solve the Van der Pol–Duffing oscillator. Stability analysis of a pair of Van der Pol oscillators with delayed self-connection, position, and velocity couplings have been investigated by Hu and Chung [12]. Qaisi [13] proposed an analytical approach based on the power series method for determining the periodic solutions of the forced undamped Duffing oscillator equation. Khan et al. [14] found an approximate solution of the force-free Duffing–Van der Pol oscillator equation using the homotopy perturbation method (HPM). Marinca and Herisanu [15] used the variational iteration method to find approximate periodic solutions of the Duffing equation with strong nonlinearity.

The Van der Pol–Duffing oscillator equation has been used in various real-life problems. A few of them are mentioned here. Hu and Wen [16] applied the Duffing oscillator for extracting the features of early mechanical failure signal. Zhihong and Shaopu [17] used the Van der Pol–Duffing oscillator equation for weak signal detection. The amplitude and phase of a weak signal have been determined by [18] using the Duffing oscillator equation. Tamaseviciute et al. [19] investigated an extremely simple analog electrical circuit simulating the double-well Duffing–Holmes oscillator equation. Weak periodic signals and machinery faults have been explained by Li and Qu [20]. Mall and Chakraverty [21] developed single-layer Hermite neural network model to handle the Van der Pol–Duffing oscillator equation.

10.1 Model Equation

The Van der Pol–Duffing oscillator equation is governed by a second-order nonlinear differential equation

$$\frac{d^2x}{dt^2} - \mu(1-x^2)\frac{dx}{dt} + \alpha x + \beta x^3 = F\cos\omega t \tag{10.1}$$

$$\text{with initial conditions } x(0) = a, x'(0) = b$$

where

x stands for displacement

μ is the damping parameter

F and ω denote the excitation amplitude and frequency of the periodic force, respectively

t is the periodic time

β is known as the phase nonlinearity parameter

Equation 10.1 reduces to the unforced Van der Pol–Duffing oscillator equation when $F = 0$.

In this chapter, we have considered single-layer simple orthogonal polynomial–based neural network (SOPNN) and Hermite neural network (HeNN) models to handle unforced and forced Van der Pol–Duffing oscillator equations, respectively.

It may be noted that the structure, formulation, error computation, and learning algorithm of the SOPNN model are already described in Sections 5.1.4.1 through 5.1.4.3, respectively. Similarly, the architecture of HeNN, and its formulation and learning algorithm are also included in Sections 5.1.3.1 and 5.1.3.2, respectively.

10.2 Unforced Van der Pol–Duffing Oscillator Equation

Here, we have considered the Van der Pol–Duffing oscillator equation without periodic force. An example problem has been solved using the single-layer SOPNN model.

Example 10.1

In this example, an unforced Van Der Pol–Duffing oscillator equation has been considered as [14].

Here, the differential equation is

$$\frac{d^2x}{dt^2}+\left(\frac{4}{3}+3x\right)\frac{dx}{dt}+\frac{1}{3}x+x^3=0$$

with initial conditions $x(0)=-0.2887$, $x'(0)=0.12$.

The related SOPNN trial solution is

$$x_\phi(t,p)=-0.2887+0.12t+t^2N(t,p)$$

In this case, we have considered 25 points in the interval [0, 10] and the first 6 simple orthogonal polynomials. We have compared SOPNN results with new homotopy perturbation method (NHPM) results [14] in Table 10.1. NHPM and SOPNN results are also compared graphically in Figure 10.1. Finally, Figure 10.2 depicts the plot of error between NHPM and SOPNN results.

10.3 Forced Van der Pol–Duffing Oscillator Equation

In this section, the forced Van der Pol–Duffing oscillator equation and the related application problem have been investigated to show the efficiency of the single-layer HeNN model.

TABLE 10.1

Comparison between NHPM and SOPNN Results (Example 10.1)

Input Data t (Time)	NHPM [14]	SOPNN
0	−0.2887	−0.2887
0.4000	−0.2456	−0.2450
0.8000	−0.2106	−0.2099
1.2000	−0.1816	−0.1831
1.6000	−0.1571	−0.1512
2.0000	−0.1363	−0.1400
2.4000	−0.1186	−0.1206
2.8000	−0.1033	−0.1056
3.2000	−0.0900	−0.0912
3.6000	−0.0786	−0.0745
4.0000	−0.0686	−0.0678
4.4000	−0.0599	−0.0592
4.8000	−0.0524	−0.0529
5.2000	−0.0458	−0.0436
5.6000	−0.0400	−0.0405
6.0000	−0.0350	−0.0351
6.4000	−0.0306	−0.0297
6.8000	−0.0268	−0.0262
7.2000	−0.0234	−0.0237
7.6000	−0.0205	−0.0208
8.0000	−0.0179	−0.0180
8.4000	−0.0157	−0.0160
8.8000	−0.0137	−0.0139
9.2000	−0.0120	−0.0115
9.6000	−0.0105	−0.0113
10.0000	−0.0092	−0.0094

Example 10.2

Here, we take the Van der Pol–Duffing oscillator equation [11] as

$$\frac{d^2x}{dt^2} + 0.2\left(x^2 - 1\right)\frac{dx}{dt} - x + x^3 = 0.53\cos t$$

subject to initial conditions $x(0) = 0.1$, $x'(0) = -0.2$.

The HeNN trial solution, in this case, is represented as

$$x_{\mathrm{He}}\left(t, p\right) = 0.1 - 0.2t + t^2 N\left(t, p\right)$$

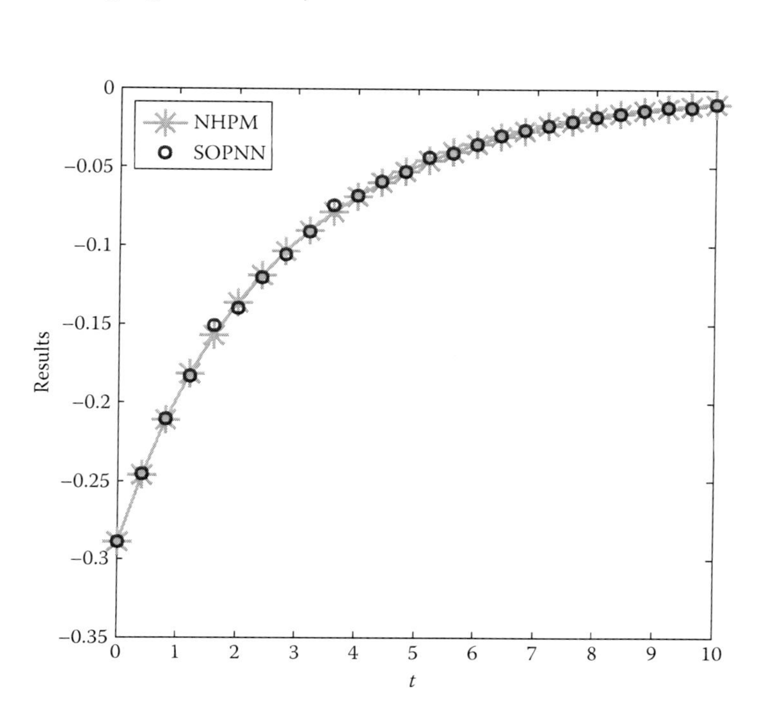

FIGURE 10.1
Plot of NHPM [14] and SOPNN results (Example 10.1).

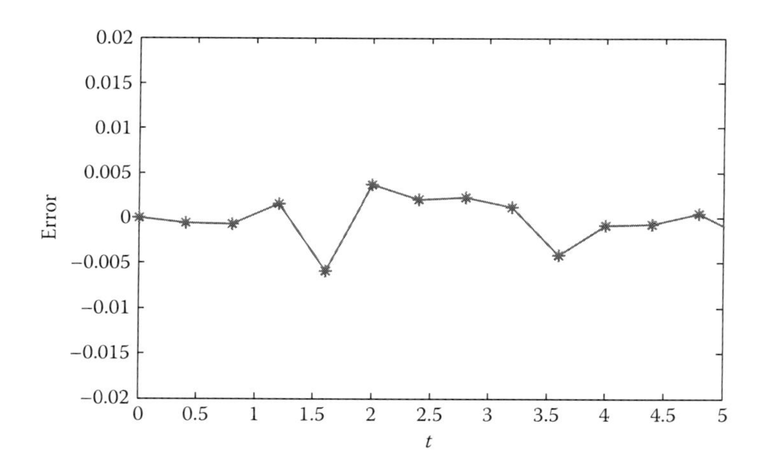

FIGURE 10.2
Error plot between NHPM and SOPNN results (Example 10.1).

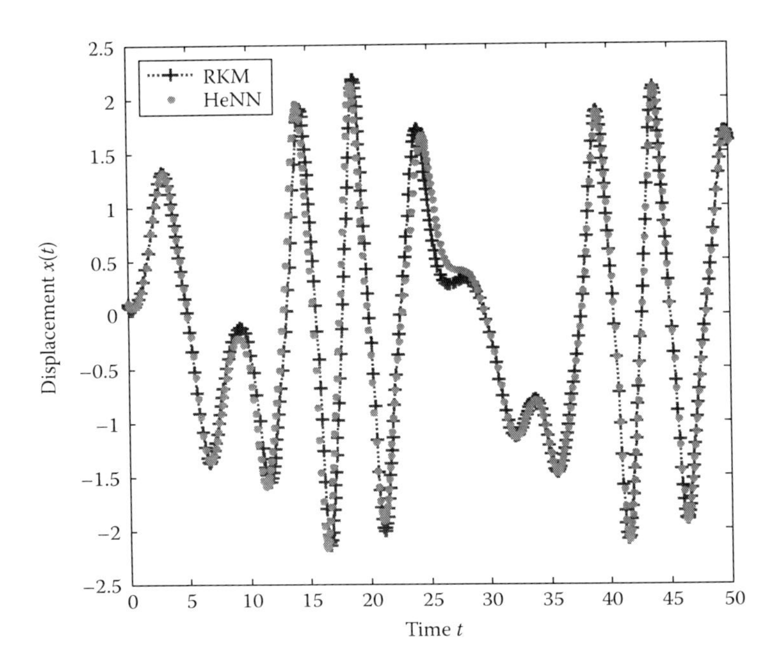

FIGURE 10.3
Plot of RKM and HeNN results (Example 10.2).

The network has been trained for 250 equidistant points in the domain, that is, from $t=0$ to $t=50$ s. for computing the results. We have considered seven weights with respect to the first seven Hermite polynomials for the present problem. Here, t denotes the periodic time and $x(t)$ is the displacement at time t. A comparison between numerical results obtained by fourth-order Runge–Kutta method (RKM) and HeNN is depicted in Figure 10.3. The phase plane diagrams, that is, plots between $x(t)$ (displacement) and $x'(t)$ (velocity), for HeNN and RKM are shown in Figures 10.4 and 10.5, respectively. Then, results for some testing points are shown in Table 10.2. This testing is done to check whether the converged HeNN can give results directly by inputting the points that were not taken during training.

Example 10.3

The Van der Pol–Duffing oscillator equation is written as [7]

$$\frac{d^2x}{dt^2} + 0.1\left(x^2 - 1\right)\frac{dx}{dt} + 0.5x + 0.5x^3 = 0.5\cos 0.79t$$

with initial conditions $x(0)=0$, $x'(0)=0$.

The HeNN trial solution may be written as

$$x_{\mathrm{He}}\left(t, p\right) = t^2 N\left(t, p\right)$$

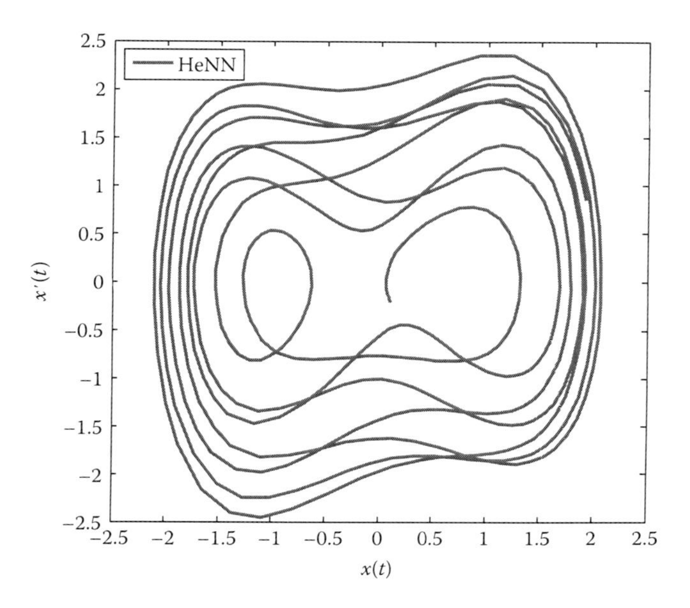

FIGURE 10.4
Phase plane plot by HeNN (Example 10.2) [21].

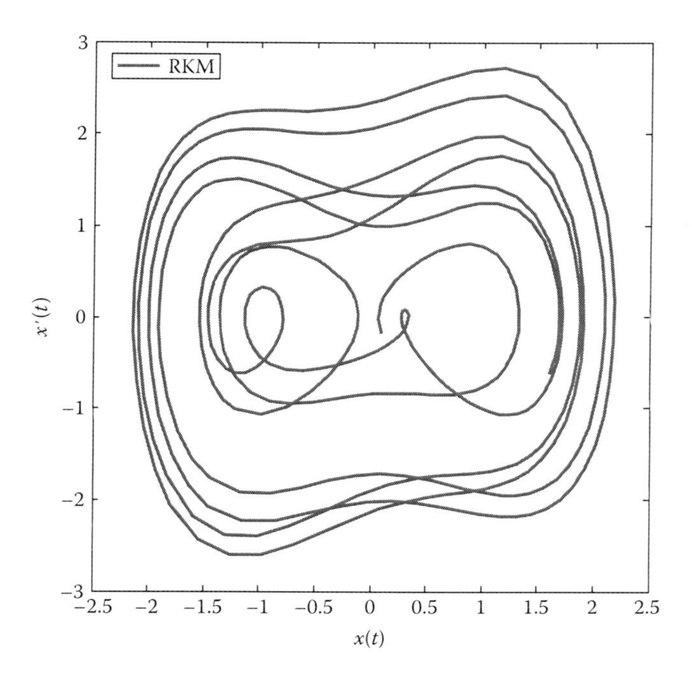

FIGURE 10.5
Phase plane plot by RKM (Example 10.2).

TABLE 10.2
RKM and HeNN Results for Testing Points (Example 10.2)

Testing Points	RKM	HeNN
1.3235	0.3267	0.3251
3.8219	0.9102	0.9092
6.1612	−1.1086	−1.1078
11.7802	−1.3496	−1.3499
18.6110	2.1206	2.1237
26.1290	0.5823	0.5810
31.4429	−0.9638	−0.9637
35.2970	−1.4238	−1.4229
43.0209	0.6988	0.6981
49.7700	1.6881	1.6879

In this case, 250 equidistant points from $t=0$ to $t=50$ s and 7 weights with respect to the first 7 Hermite polynomials have been considered for the present problem. RKM and HeNN results are compared in Figure 10.6. Figures 10.7 and 10.8 show the phase plane plots obtained by HeNN and RKM, respectively.

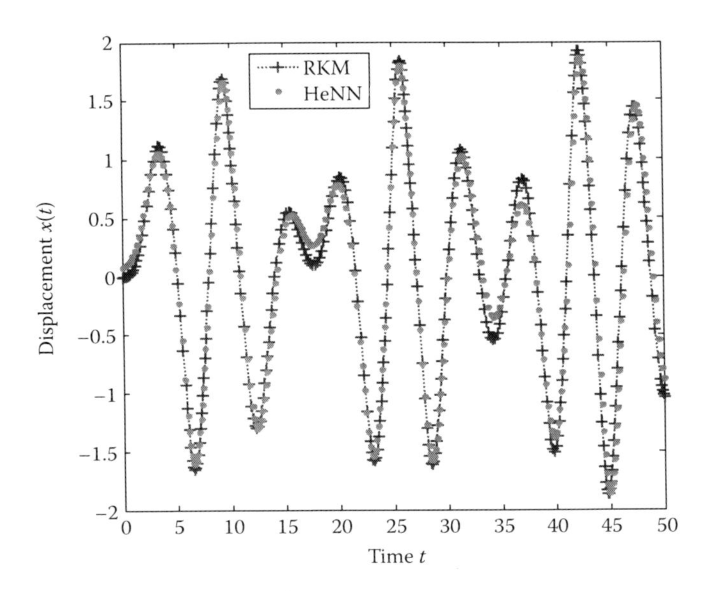

FIGURE 10.6
Plot of RKM and HeNN results (Example 10.3).

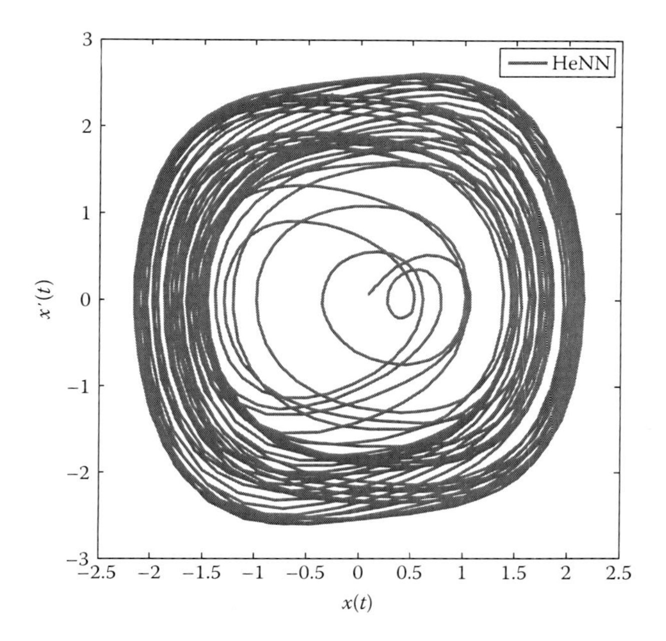

FIGURE 10.7
Phase plane plot by HeNN (Example 10.3) [21].

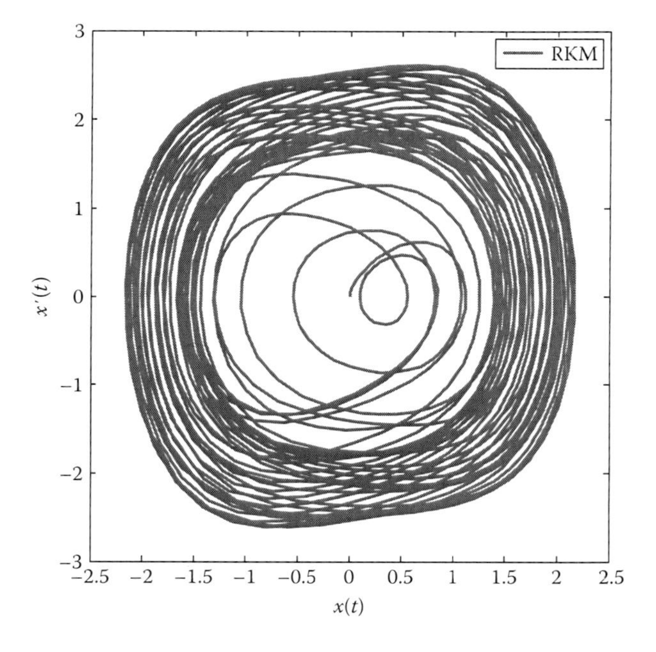

FIGURE 10.8
Phase plane plot by RKM (Example 10.3).

Example 10.4

Finally, the Van der Pol–Duffing oscillator equation applied for weak signal detection can be written as [17]

$$\frac{d^2x}{dt^2} - \mu\left(x^2 - 1\right)\frac{dx}{dt} + x + \alpha x^3 = F\cos\omega t$$

subject to the initial conditions $x(0) = 0.1$, $x'(0) = 0.1$, with the parameters $\mu = 5$, $\alpha = 0.01$, $w = 2.463$, and $F = 4.9$.

The HeNN trial solution in this case is

$$x_{\mathrm{He}}\left(t,p\right) = 0.1 + 0.1t + t^2 N\left(t,p\right)$$

Again, the network is trained for total time t = 300 s and h = 0.5. It may be noted that [17] have solved the problem by fourth-order RKM. The time series plots by RKM [17] and HeNN are depicted in Figures 10.9 and 10.10, respectively. Last, the phase plane plots obtained by using RKM [17] and HeNN [21] are given in Figures 10.11 and 10.12, respectively.

It may be noted that as the amplitude of force F varies from small to large, the Van der Pol–Duffing system varies from the chaotic state to the periodic state. The results show that the orbits maintain the chaotic state. The detected signal can be viewed as a perturbation of the main sinusoidal deriving force $F\cos\omega t$. The noise can only affect the local trajectory on the phase plane diagram without causing any phase transition.

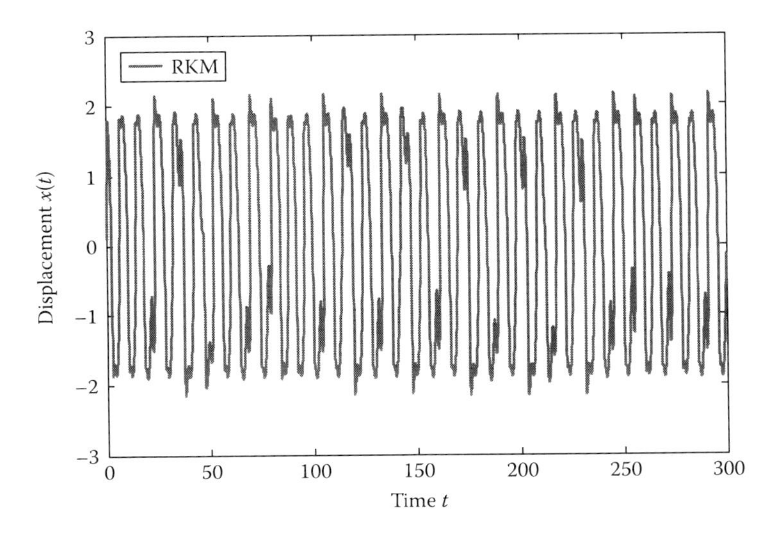

FIGURE 10.9
Time series plot of RKM [17] (Example 10.4).

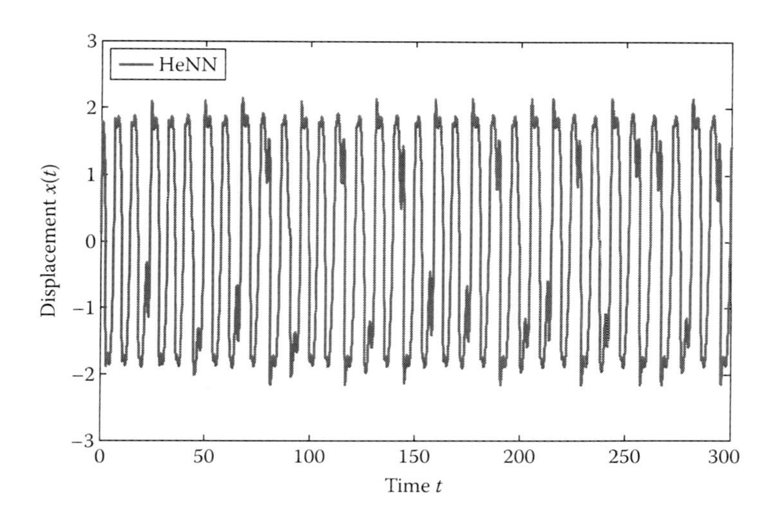

FIGURE 10.10
Time series plot of HeNN (Example 10.4) [21].

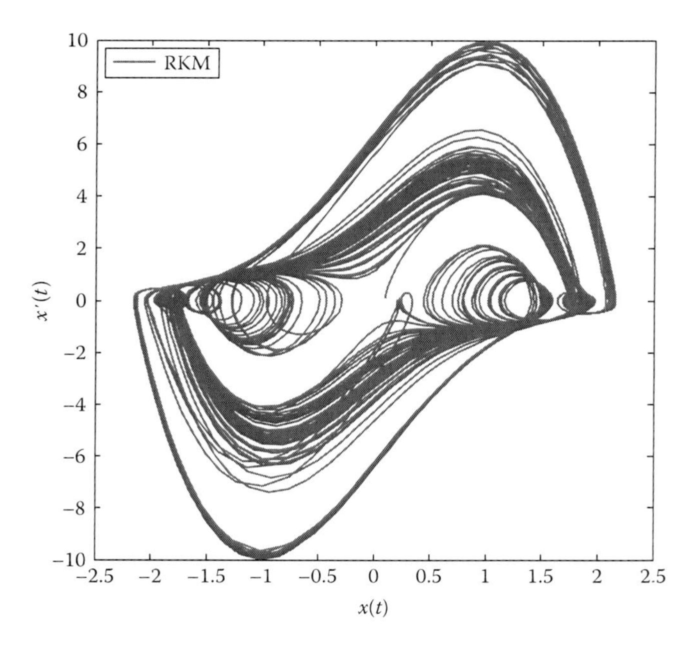

FIGURE 10.11
Phase plane plot by RKM [17] (Example 10.4).

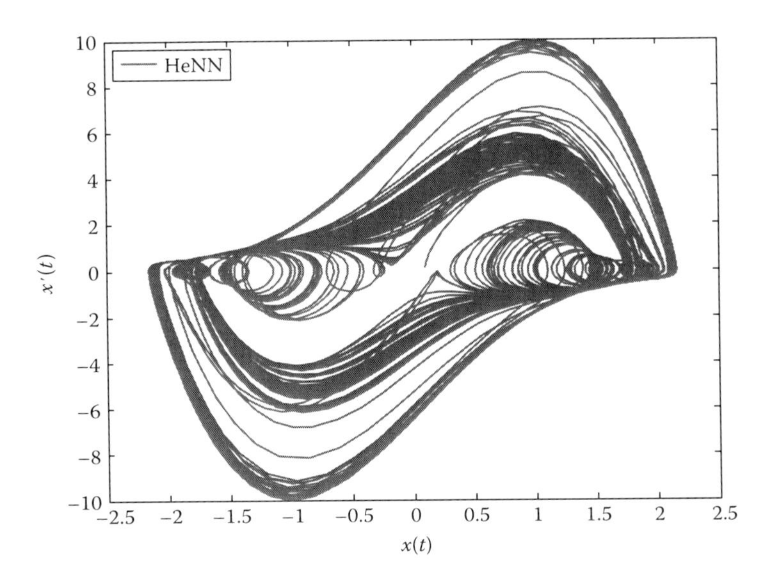

FIGURE 10.12
Phase plane plot by HeNN (Example 10.4).

Again, one finds an excellent agreement of results between RKM [17] and the present (HeNN) method for time series results (viz. Figures 10.9 and 10.10) and phase plane plots (Figures 10.11 and 10.12).

References

1. J. Guckenheimer and P. Holmes. *Nonlinear Oscillations, Dynamical Systems and Bifurcations of Vector Fields*. Springer-Verlag, New York, 1983.
2. M. Tsatssos. Theoretical and numerical study of the Van der Pol equation, Dissertation, Thessaloniki, Greece, 2006.
3. S.S. Motsa and P. Sibanda. A note on the solutions of the Van der Pol and Duffing equations using a linearization method. *Mathematical Problems in Engineering*, 2012: 1–10, July 2012.
4. M.R. Akbari, D.D. Ganji, A. Majidian, and A.R. Ahmadi. Solving nonlinear differential equations of Vander Pol, Rayleigh and Duffing by AGM. *Frontiers of Mechanical Engineering*, 9(2): 177–190, June 2014.
5. S. Nourazar and A. Mirzabeigy. Approximate solution for nonlinear Duffing oscillator with damping effect using the modified differential transform method. *Scientia Iranica*, 20(2): 364–368, April 2013.
6. S. Mukherjee, B. Roy, and S. Dutta. Solutions of the Duffing–van der Pol oscillator equation by a differential transform method. *Physica Scripta*, 83: 1–12, December 2010.

7. A. Kimiaeifar, A.R. Saidi, G.H. Bagheri, M. Rahimpour, and D.G. Domairr. Analytical solution for Van der Pol–Duffing oscillators. *Chaos, Solutions and Fractals*, 42(5): 2660–2666, December 2009.
8. A.N. Njah and U.E. Vincent. Chaos synchronization between single and double wells Duffing Van der Pol oscillators using active control. *Chaos, Solitons and Fractals*, 37(5): 1356–1361, September 2008.
9. H. Molaei and S. Kheybari. A numerical solution of classical Van der Pol-Duffing oscillator by He's parameter-expansion method. *International Journal of Contemporary Mathematical Sciences*, 8(15): 709–714, 2013.
10. N.H. Sweilam and R.F. Al-Bar. Implementation of the parameter-expansion method for the coupled van der Pol oscillators. *International Journal of Nonlinear Sciences and Numerical Simulation*, 10(2): 259–264, 2009.
11. C. Zhang and Y. Zeng. A simple numerical method For Van der Pol-Duffing Oscillator Equation. *International Conference on Mechatronics, Control and Electronic Engineering*, Shenyang, China, Atlantis Press, pp. 476–480, September 2014.
12. K. Hu and K.W. Chung. On the stability analysis of a pair of Van der Pol oscillators with delayed self-connection position and velocity couplings. *AIP Advances*, 3: 112–118, 2013.
13. M.I. Qaisi. Analytical solution of the forced Duffing oscillator. *Journal of Sound and Vibration*, 194(4): 513–520, July 1996.
14. N.A. Khan, M. Jamil, S.A. Ali, and N.A. Khan. Solutions of the force-free Duffing-van der pol oscillator equation. *International Journal of Differential Equations*, 2011: 1–9, August 2011.
15. V. Marinca and N. Herisanu. Periodic solutions of Duffing equation with strong non-linearity. *Chaos, Solitons and Fractals*, 37(1): 144–149, July 2008.
16. N.Q. Hu and X.S. Wen. The application of Duffing oscillator in characteristic signal detection of early fault. *Journal of Sound and Vibration*, 268(5): 917–931, December 2003.
17. Z. Zhihong and Y. Shaopu. Application of van der Pol–Duffing oscillator in weak signal detection. *Computers and Electrical Engineering*, 41: 1–8, January 2015.
18. G. Wang, W. Zheng, and S. He. Estimation of amplitude and phase of a weak signal by using the property of sensitive dependence on initial conditions of a nonlinear oscillator. *Signal Processing*, 82(1): 103–115, January 2002.
19. E. Tamaseviciute, A. Tamasevicius, G. Mykolaitis, and E. Lindberg. Analogue electrical circuit for simulation of the Duffing–Holmes Equation. *Nonlinear Analysis: Modelling and Control*, 13(2): 241–252, June 2008.
20. C. Li and L. Qu. Applications of chaotic oscillator in machinery fault diagnosis. *Mechanical Systems and Signal Processing*, 21(1): 257–269, January 2007.
21. S. Mall and S. Chakraverty. Hermite functional link neural network for solving the Van der Pol-Duffing oscillator equation. *Neural Computation*, 28(8): 1574–1598, July 2016.

Index

H

I

L

M

N

O

P

R

S